Life Science

Written by
Pamela Jennett

Editor: Collene Dobelmann
Illustrator: Darcy Tom
Designer/Production: Moonhee Pak/Andrea Ables
Cover Designer: Barbara Peterson
Art Director: Tom Cochrane
Project Director: Carolea Williams

Table of Contents

Introduction

Each book in the *Power Practice*™ series contains over 100 ready-to-use activity pages to provide students with skill practice. The fun activities can be used to supplement and enhance what you are teaching in your classroom. Give an activity page to students as independent class work, or send the pages home as homework to reinforce skills taught in class. An answer key is provided for quick reference.

The practical activities, charts, diagrams, and definition pages in *Life Science* supplement and enrich classroom teaching to enhance students' understanding of vocabulary, functions, and processes fundamental to living organisms. This book features the following topics:
• Kingdom Animalia
• Kingdom Monera
• Kingdom Protista
• Kingdom Fungi
• Kingdom Plantae
• photosynthesis
• metamorphosis
• ecology
• Earth's biomes
• root systems
• and more!

Use these ready-to-go activities to "recharge" skill review and give students the power to succeed!

Name _____   Date _____

Life Processes

What is the difference between a living and nonliving thing? A living thing must carry out six basic life processes: get energy, use energy, get rid of waste, reproduce, grow, and respond to change. Nonliving things may be able to do one or two of these processes, but they cannot do all of them and are therefore nonliving. Use the phrases in the word box to label each example shown in the illustrations.

getting energy	using energy	getting rid of waste
growing	reproducing	responding to change

1 _____

2 _____

3 _____

4 _____

BANG!

5 _____

6 _____

7 _____

8 _____

9 _____

Life Science © 2004 Creative Teaching Press

The Carbon Dioxide-Oxygen Cycle

Matter is continuously cycled between the living and nonliving parts of an ecosystem and between ecosystems. Matter is recycled, no new matter is added to the earth and none is lost. One example of this is in the carbon dioxide-oxygen cycle. Match each term in the word box with its description.

> carbon dioxide   oxygen   photosynthesis   marine algae   decomposers
> producers   consumers aerobic   geologic activity fossil fuels

1 _____ Carbon is present in Earth's atmosphere in the form of this gas.

2 _____ The world's oceans hold most of the carbon in a dissolved form. These organisms use the carbon and release oxygen back into the atmosphere.

3 _____ Plants, also called this, use carbon dioxide to make their own food.

4 _____ This process, used by producers, releases oxygen into the atmosphere as a byproduct.

5 _____ These organisms cycle carbon through their bodies through the foods they eat. After they die and decompose, carbon is released back into the soil and atmosphere.

6 _____ The burning of these has put more carbon back into the atmosphere than can be cycled naturally.

7 _____ These organisms feed off of dead material and release the carbon back into the cycle.

8 _____ This type of respiration uses oxygen and produces carbon dioxide as a byproduct.

9 _____ Examples of this include volcanic eruptions and weathering of limestone rock, both of which release carbon into the atmosphere.

10 _____ The respiration of consumers uses this gas and releases carbon dioxide as a byproduct.

Name _____ Date _____

The Nitrogen Cycle

Another important cycle is the nitrogen cycle. Nitrogen is necessary for life. It is found in all proteins that cells use for growth. The amount of nitrogen stays the same as it is cycled through living and nonliving things. Match each term in the word box to its description.

lightning	leaching	ammonia	legumes	amino acids
bacteria	atmosphere	decomposition	nitrification	animals

1 _____ Nitrogen is a building block of these, a component of protein in all living things.

2 _____ 80% of this consists of nitrogen, making it the largest reservoir of this element on Earth.

3 _____ Only certain bacteria, volcanic action, and this can break down nitrogen in the air and convert it into a form that enters food webs.

4 _____ Nitrogen is fixed into the soil for plants to use through the activities of this.

5 _____ Plants use nitrogen in the soil as they grow. Directly or indirectly, plants are the only nitrogen source for these.

6 _____ During this process bacteria and fungi break down wastes and remains of organisms into ammonia.

7 _____ This process occurs when water in the soil moves out of an area, taking the dissolved nitrogen and other nutrients with it.

8 _____ This plant crop is able to fix nitrogen from the atmosphere into the soil.

9 _____ During this process bacteria convert nitrogen in the soil and release it back into the atmosphere.

10 _____ When plants and animals die, their nitrogen compounds are broken down into this.

Name _____ Date _____

The Five Kingdoms

Scientists classify all living things into five different groups. These groups are called kingdoms. Use the terms in the word box to label the kingdom and common name for each group.

Animalia	Plantae	Monera	Fungi	Protista
plants	bacteria	single-celled organisms		animals
molds, mushrooms, lichen				

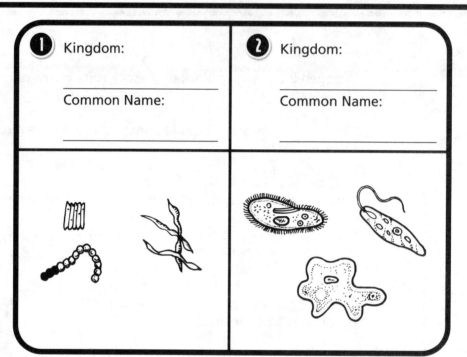

1 Kingdom: _____

Common Name:
_____

2 Kingdom: _____

Common Name:
_____

3 Kingdom: _____

Common Name:
_____

4 Kingdom: _____

Common Name:
_____

5 Kingdom: _____

Common Name:
_____

Name _____ Date _____

Classification of Living Things

Within a kingdom, the species are connected in some way. In the two largest kingdoms, Plantae and Animalia, the diversity among the species can be great. Each kingdom can be further divided into smaller and more precise categories until only a single species is represented. Use the terms in the word box to complete the chart.

conifers	bacillis	lichens	bread molds	monocots
ferns	mushrooms	salmonella	mammals	reptiles
truffles	amoeba	streptococcus	paramecium	birds
yeast	protozoa	euglena	amphibians	dicots
insects	algae	spirochetes	lactobacillus	horsetails

Kingdom Monera	Kingdom Protista
Examples:	Examples:
_____	_____
_____	_____
_____	_____
_____	_____
_____	_____

Kingdom Fungi	Kingdom Plantae	Kingdom Animalia
Examples:	Examples:	Examples:
_____	_____	_____
_____	_____	_____
_____	_____	_____
_____	_____	_____
_____	_____	_____

Plant or Animal?

Two of the largest groups of all living things are the plant and animal kingdoms. These two kingdoms have distinct characteristics while having others in common. Use the phrases in the word box to complete the chart. Some phrases are used more than once.

living organisms	formed from cells	cells have chlorophyll
cells have no chlorophyll	makes own food	moves from place to place
obtains food from outside sources	has limited movement	reproduces its own kind
depends on sun's energy		

Plant	Animal

_____

_____

_____

_____

_____

_____

_____

_____

_____

_____

The Structure of Bacteria

Kingdom Monera includes the bacteria. These are single-celled organisms that do not have a nucleus. Bacteria are the most common organisms on earth and are connected in some manner to all other organisms. Bacteria are found in the soil, in deserts and oceans, and on your skin. Use the terms in the word box to label the diagram.

bacterium flagellum	cell wall	DNA	pilus
capsule	plasma membrane	cytoplasm	ribosomes

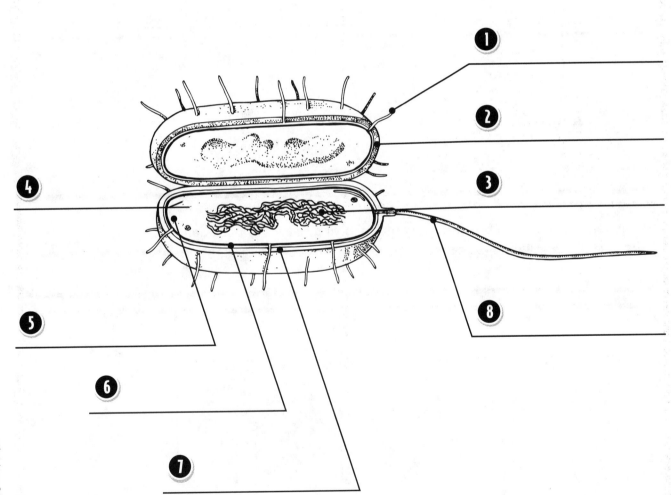

Name _____ Date _____

Bacterial Shapes

Bacteria are often classified as cocci, bacilli, or spirochetes, based on their physical appearance. Use the terms in the word box to label the classification and description of each bacteria.

spherical cocci	rod-shaped spirochetes	corkscrew-shaped bacilli

1 Type: _____

Description:

_____

2 Type: _____

Description:

_____

3 Type: _____

Description:

_____

Use the terms in the word box to classify the following species of bacteria.

cocci	spirochetes	bacilli

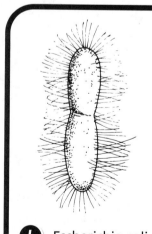

4 Escherichia coli

_____

5 Bacillis

_____

6 Streptococcus

_____

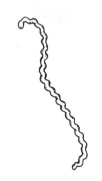

7 Leptospirilla

_____

Classifications of Bacteria

Scientists classify bacteria according to what they need in order to survive and their effects on other organisms. Match each term in the word box to its definition.

autotrophs   anaerobic   cyanobacteria   chemoautotrophs
pathogenic   heterotrophs   beneficial   aerobic

1 _____ These bacteria require oxygen in order to live.

2 _____ These bacteria are organisms that get carbon from carbon dioxide gas.

3 _____ These bacteria can survive without the presence of oxygen.

4 _____ These bacteria are needed for body functions or to keep us healthy.

5 _____ Also known as blue-green algae, these bacteria contain chlorophyll and get energy from sunlight.

6 _____ These bacteria get energy from other chemicals.

7 _____ These organisms get energy by ingesting organic molecules from organisms or by preying on other bacteria.

8 _____ These bacteria cause disease or illness.

Name _____ Date _____

What It Takes to Be a Protistan

The members of the Kingdom Protista are classified because of what they are not. Protists do not have all the traits of bacteria, fungi, plants, or animals and are therefore classified in their own kingdom. The euglena is one example of a protist. Use the terms in the word box to label the diagram.

long flagellum	contractile vacuole	chloroplast
pellicle	mitochondrion	Golgi body
nucleus	endoplasmic reticulum	light-sensitive spot

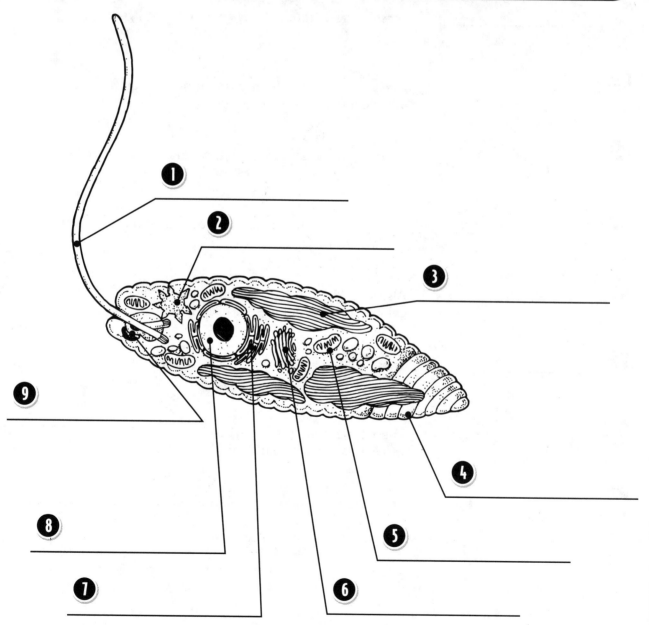

Life Science © 2004 Creative Teaching Press

Name _____ Date _____

The Paramecium

The paramecium is another example of a protist. It is able to move through water through the movement of cilia, or hair-like structures along its cell body. Use the terms in the word box to label the diagram. Some words may be used twice.

> contractile vacuole   micronucleus   macronucleus
> gullet   trichocysts   food vacuole
> cilia

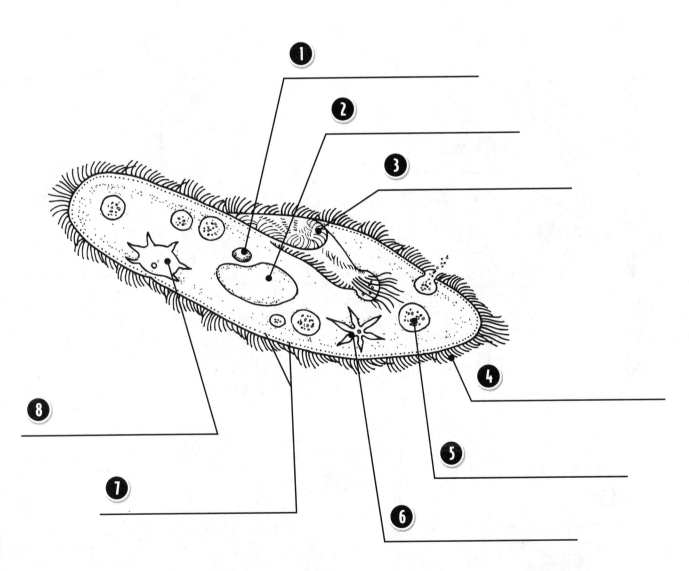

Moving Like an Amoeba

Amoebas are another member of the Kingdom Protista. These living organisms move or capture prey by sending out a pseudopod or "false foot," an extension of their cell body. Amoebas reproduce by dividing. Use the terms in the word box to label the reproducing amoebas in the diagram. Some words may be used twice.

| food vacuole | nucleus | water vacuole |
| false foot | cell membrane | cytoplasm |

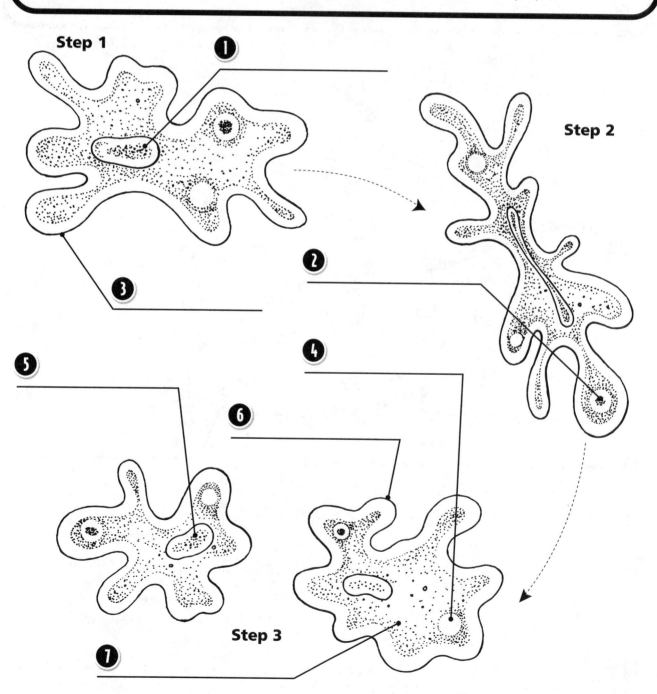

Name _____ Date _____

Red, Brown, and Green Algae

Red, brown, and green algae are all members of the Kingdom Protista. Red algae are mostly found in warm seas. They get their color from the pigments in their cells. They are found at deep depths underwater, and these pigments pick up the light rays able to reach to that depth. Green algae are most similar to plants and may be a close relative. They are found mostly in freshwater. Brown algae live in cool marine waters. Giant kelp is familiar example of brown algae. Use the terms in the word box to label the diagram of brown algae.

| holdfast | bladder | blades | stipe |

1 _____

2 _____

3 _____

4 _____

Match each term in the word box above to its definition.

5 _____ This structure anchors the algae to rocks or other surfaces.

6 _____ Similar to a stem, these tube-like structures conduct food throughout the algae.

7 _____ This hollow, gas-filled structure gives the algae buoyancy and allows them to float.

8 _____ These leaflike parts spread out in the water, allowing the algae to collect the sunlight.

Name _____ Date _____

A Fungus Among Us

The Kingdom Fungi are a group of organisms that contain chitin within their cell walls. They lack chlorophyll and therefore cannot make their own food. Use the information in the descriptions to label the type of fungus shown in each illustration.

Mushroom: These have a stalk and cap. Under the cap are gills where the spores are produced.

Crusty Lichen: Grows on rock and the ground. They look almost as though something crusty has been painted onto the surface of a rock.

Shrubby Lichen: Sometimes mistakenly called a "moss," this lichen looks like a hairy plant growing from tree branches.

Mold: Looking like white, green, gray, or black fur, it grows on food and other surfaces that remain moist and at certain temperatures.

Yeast: So small, these can only be seen when grouped together. They form a silvery bloom on fruit.

Boletus: Has a cap and stalk, but instead of gills there is soft tissue containing a lot of tiny holes.

Bracket Fungi: Grows on trees, often sticking out like little ruffles or shelves.

Puffballs: Little balls filled with spores that burst open to release spores when hit by a raindrop.

Life Science © 2004 Creative Teaching Press

The Nature of Fungi

The members of the fungi kingdom have unique characteristics that allow them to survive. Match each term in the word box to its definitions.

mold	parasite	fungi	fungus	lichen
yeast	penicillin	budding	toadstool	chlorophyll

1 _____ This mold is useful. It inhibits the growth of bacteria.

2 _____ This furlike fungi thrives on organic substances such as fruit and leather. With enough moisture it can disintegrate these substances rapidly.

3 _____ This is the form of the word meaning just one.

4 _____ A member of the fungi kingdom, this organism coexists with algae to be able to survive in harsh locations.

5 _____ This is the process by which yeast cells can reproduce new cells.

6 _____ Certain types are useful in the fermentation of wine or the baking of bread. It is a single-celled fungi.

7 _____ The members of this group lack the ability to make their own food and contain chitin within their cell walls. It is one of the five kingdoms of living things.

8 _____ This is another word for a mushroom.

9 _____ Fungi lack this substance and therefore have developed unique ways to absorb food material.

10 _____ Some fungi act as this to plants, animals, or other fungi and can destroy the host organism.

Parts of a Mushroom

Mushrooms are probably the most well-known member of the fungi kingdom. Mushrooms are decomposers; they get their nutrients by breaking down matter of organisms that are no longer alive. Use the terms in the word box to label the diagram.

cap   gills   ring   stalk   mycelium   membrane

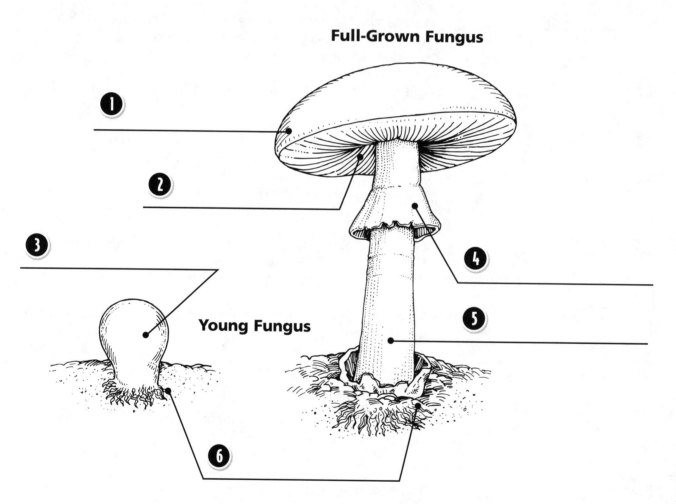

Full-Grown Fungus

Young Fungus

➊
➋
➌
➍
➎
➏

Growing Up Mushroom

When a spore falls to the ground, and conditions are right, it starts to grow. First it makes an underground network of threads called spawn. From this spawn the mushroom grows. Use the terms in the word box to label the growth stages of a mushroom.

| pinhead stage | full-grown | half-grown | button stage |

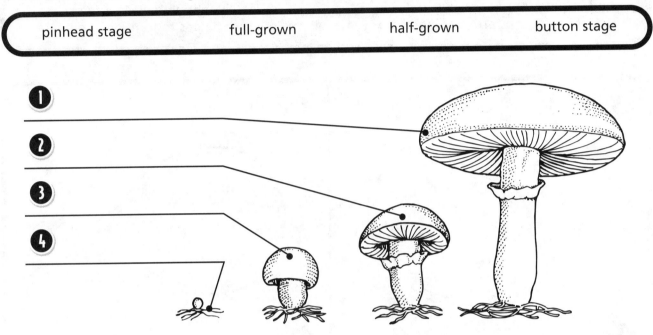

1

2

3

4

As a mushroom grows, small changes take place. Use the terms in the word box to label the parts of the mushroom as it grows.

| spawn   gills   ring   gills inside of veil   veil   veil broken to form ring |

5

6

7

8

9

10

Dividing to Multiply

Use the phrases in the word box to identify each step in the diagrams.

> bud forms as nucleus begins to divide
> nucleus moves toward dividing end of yeast cell
> cells fully split with nucleus in each cell
> cell wall begins to form new cell

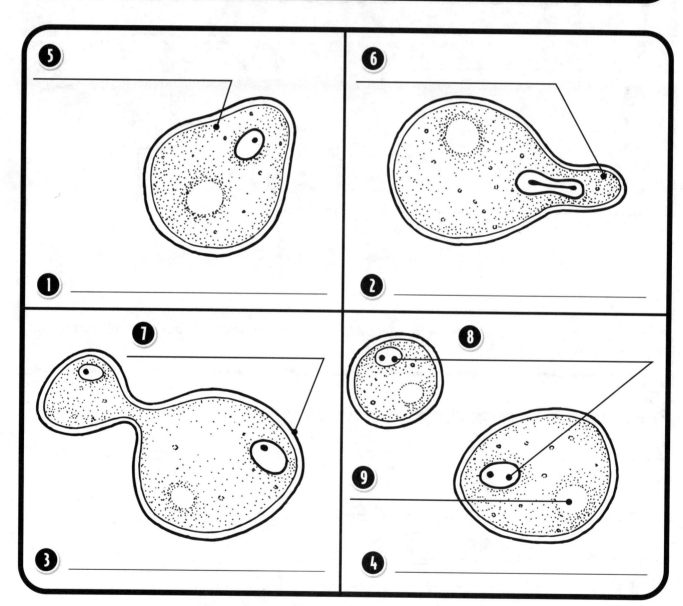

Use the terms in the word box to label the parts of the yeast cells in the diagrams above.

> bud   cell wall   cytoplasm   nuclei   vacuole

Name _____ Date _____

The World of Plants

The plant kingdom is made up of organisms that contain chlorophyll and have rigid cell walls made of cellulose. Use the definitions below to label each type of plant shown in the illustrations.

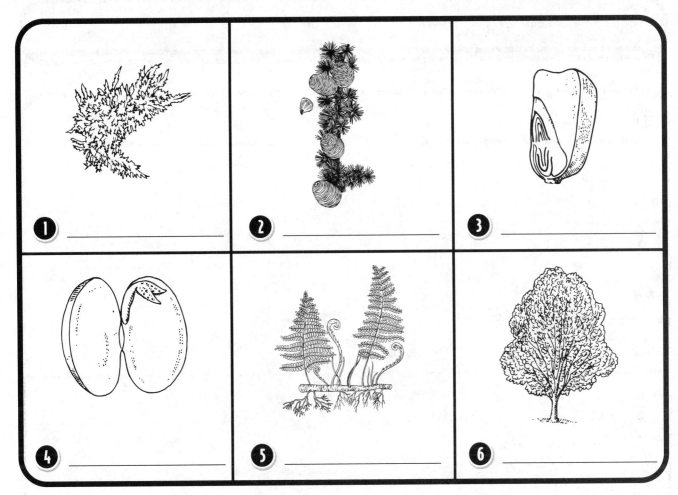

Monocot: A flowering plant with seeds that have only one cotyledon, flower parts that occur in threes, and leaves that are generally parallel-veined.

Deciduous Tree: A tree that loses its leaves at a particular season.

Dicot: A flowering plant with seeds that have two cotyledons, flower parts that occur in fours or fives, and leaves that are net-veined.

Moss: A small, green plant that is nonvascular, meaning it lacks a system of tubes through which water is transported through the plant.

Conifer: A mostly evergreen tree or shrub with needlelike leaves and seeds in cones.

Fern: A vascular plant that lacks flowers or seeds but reproduces with spores.

Name _____   Date _____

A Typical Plant Cell

Plant cells have basic structures in common, even though plant cells are as varied as the plants themselves. Each individual plant cell is partly self-sufficient, carrying on processes contained within the cell membrane. A plant cell differs from an animal cell because it contains chloroplasts and has a cell wall made of cellulose. Use the terms in the word box to label the diagram.

> cytoplasm   chloroplast   vacuole
> endoplasmic reticulum   ribosomes   nucleus
> mitochondrion   cell wall   cell membrane

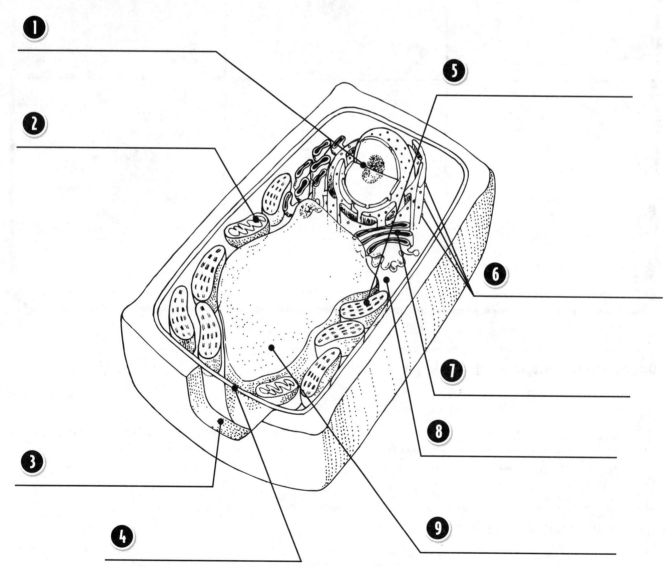

Name _____ Date _____

Functions within a Plant Cell

Each of the structures, or organelles, within a plant cell serves a specific purpose. Match each term in the word box to its definition.

> cytoplasm   chloroplast   vacuole
> endoplasmic reticulum   ribosome   nucleus
> mitochondrion   cell wall   organelle
> cell membrane

1 _____ This is the tough, nonliving outer layer of each plant cell. It gives the cell shape, strength, and support.

2 _____ This is a structure that stores water and helps keep the plant from wilting.

3 _____ This is a structure that contains chlorophyll, giving the plant its green color, and traps energy from sunlight.

4 _____ This is a structure that moves material throughout the cell.

5 _____ This is a substance that fills most of the cell outside the nucleus and contains the other organelles.

6 _____ This is any tiny structure in the cytoplasm of the cell that performs a special job.

7 _____ This is an organelle that puts together proteins for the cell.

8 _____ This is a structure where food and oxygen react to release energy.

9 _____ This acts as the control center for the cell.

10 _____ This is a layer that holds the parts of the cell together and controls movement of materials into and out of the cell.

Name _____ Date _____

Photosynthesis

Photosynthesis is a process unique to the plant world. *Photo* means "light" and *synthesis* means "putting together." *Photosynthesis*, therefore, means "putting together with light." Through this process, plants are able to make their own food. Use the terms in the word box to complete each sentence.

photosynthesis	oxygen	carbohydrates	water
heterotrophs	carbon dioxide	sunlight	chlorophyll
chloroplasts	glucose	cellulose	leaves
autotrophs			

1 Water, carbon dioxide, and _____ are the three ingredients needed for photosynthesis to occur.

2 Through photosynthesis, plants convert these ingredients into _____, a food used by the plant.

3 _____ is the material in green plant cells that traps energy from the sun.

4 The plant takes in a gas called _____ from the air.

5 Chlorophyll is found in the _____, structures within the cell where photosynthesis will take place.

6 _____ is a material the plant takes up through its roots and stems.

7 During photosynthesis, _____ is a waste product released by the plant into the air.

8 Plants produce more glucose than they need. The excess glucose is stored by the plant as _____ or starches.

9 Plants also change glucose into _____, the structural material used in their cell walls.

10 While chlorophyll is found in most aboveground parts of green plants, most photosynthesis takes place in a plant's _____.

11 In many regions, there is not enough sunlight or water during winter for _____ to occur.

12 Most plants are also called _____ because they are able to produce their own food.

13 Most animals are called _____ because they obtain energy from other plants or animals.

Life Science © 2004 Creative Teaching Press

Mosses

Mosses grow on soil, rocks, the bark of trees, and in bogs and shallow streams. They are also known as bryophytes, meaning that they do not have a vascular system to carry nutrients and water throughout the plant. Mosses lack true roots and do not reproduce by seeds but rather by spores. Use the terms in the word box to label the diagram.

rhizoid	gametophyte	sporophyte
leaves	stems	foot
seta	calyptra	capsules

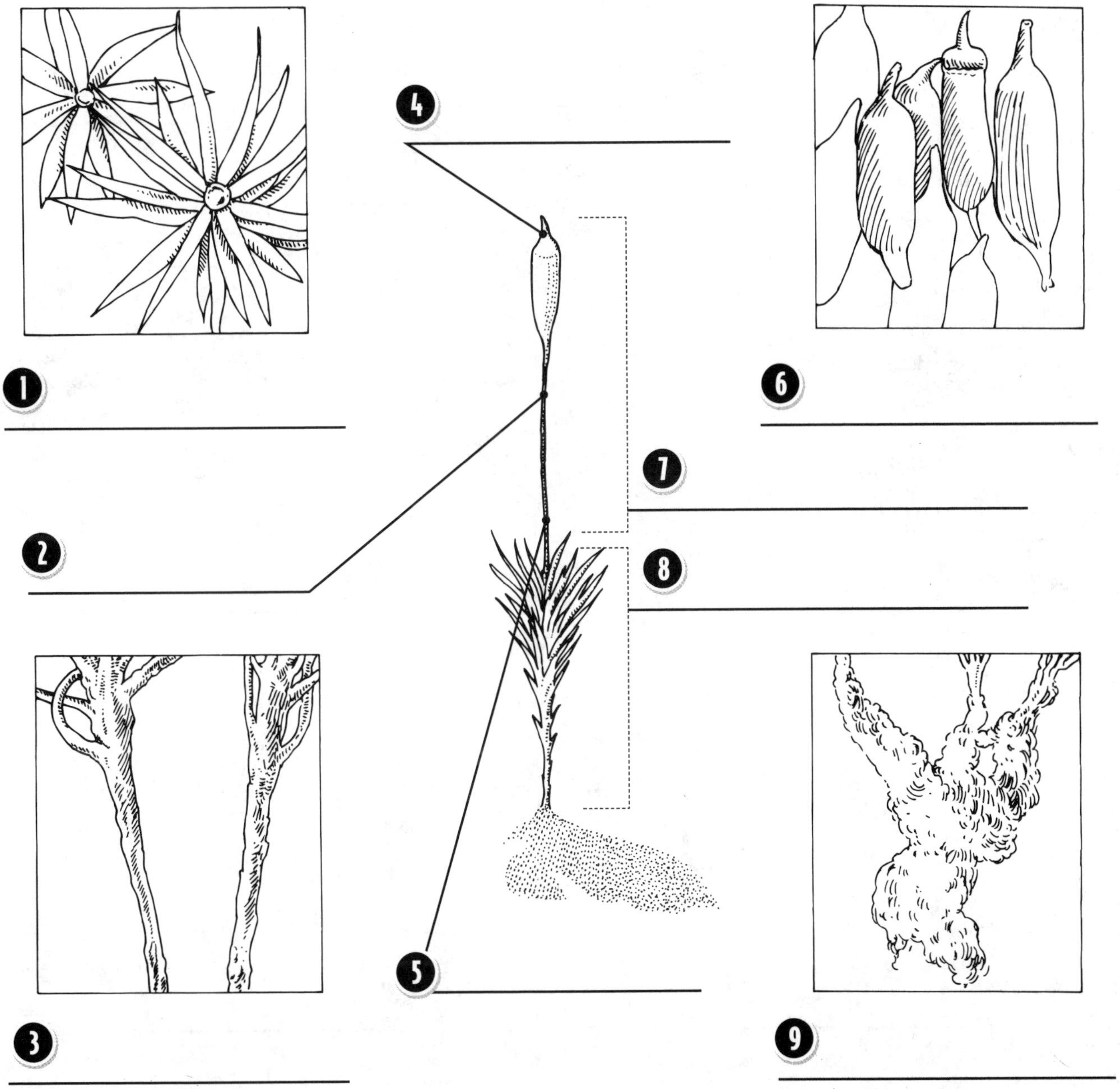

❶ _____

❷ _____

❸ _____

❹ _____

❺ _____

❻ _____

❼ _____

❽ _____

❾ _____

Ferns

Ferns are flowerless, seedless plants that reproduce by forming spores. The spores grow in structures located on the underside of the fronds. Ferns are vascular plants, meaning that a series of tubes within the stems and leaves carry nutrients and water throughout the plant. Use the terms in the word box to label the diagram.

frond	fiddlehead	rhizome	root
spores	leaflets	sorus	

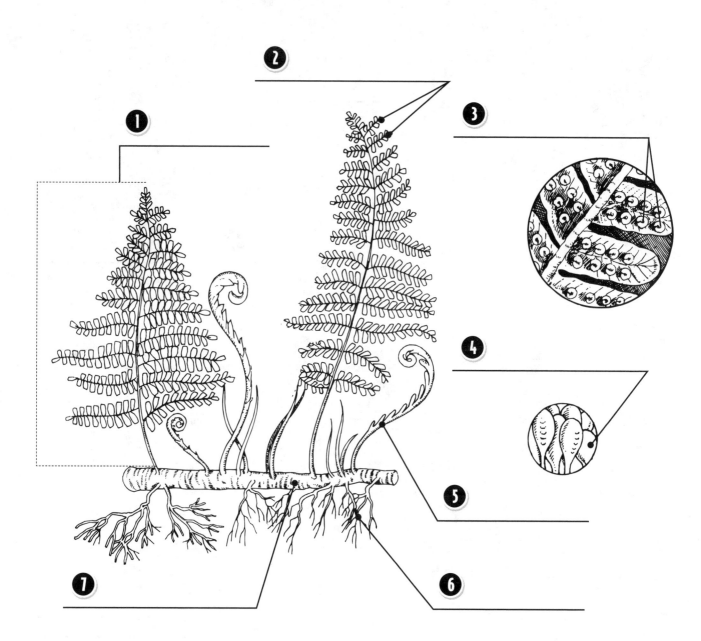

Horsetails

Of all existing plants, the horsetails may be the oldest. By examining fossil records of these plants, it appears that they have changed very little since they first appeared on Earth. Their stems are reinforced with ribs of silica, giving them a sandpaper feel. Pioneers of the American West gathered horsetails to use as pot scrubbers. Use the terms in the word box to label the diagram.

| hollow stems | scalelike leaves | rhizomes | ribs | spore-bearing structures |

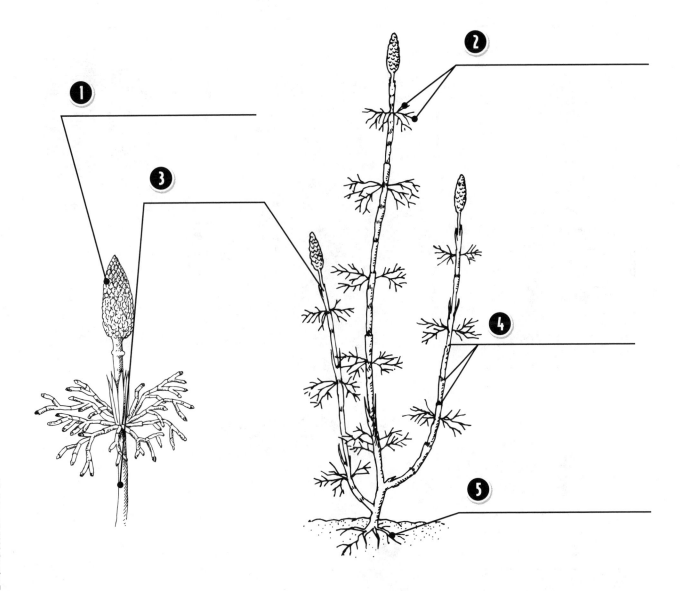

Gymnosperms

Gymnosperm means "naked seed." Gymnosperms are plants that make their seeds on exposed parts like cones. Conifers such as the pine tree or the giant sequoia are gymnosperms. So are gingko trees and sago palms. Use the terms in the word box to label the diagram.

cone	seed	needles
spores	trunk	roots

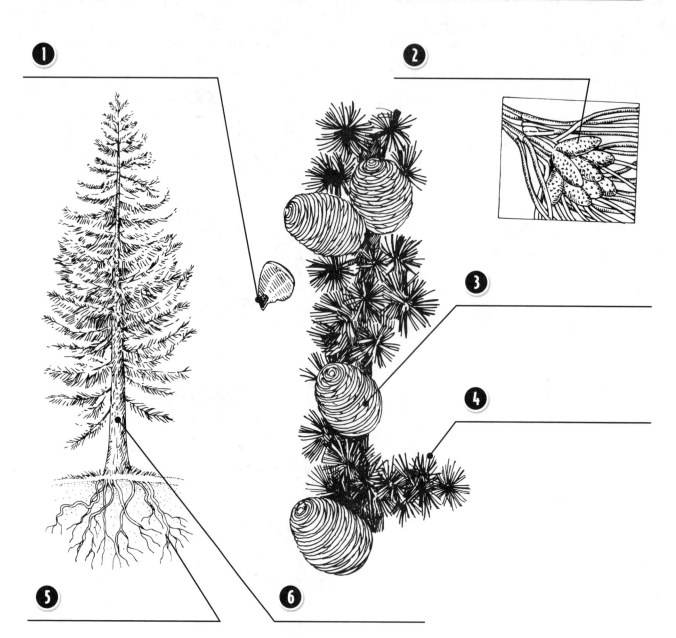

Life Science © 2004 Creative Teaching Press

Conifer Needles, Scales, and Cones

Conifers are also known as evergreens because most often these trees stay green year-round. The leaves on conifers are known as needles or scales. Thin and sharp, the leathery surface allows them to weather the elements. The seeds of conifers are protected by hard cones. The scales of a cone loosen as the seeds ripen. Use the terms in the word box to label the diagrams.

cone scale   mature cone   young cone
shell around seed kernel   seed kernel   needles in pairs
needles in threes   needles in fives   flat needles
scalelike needles

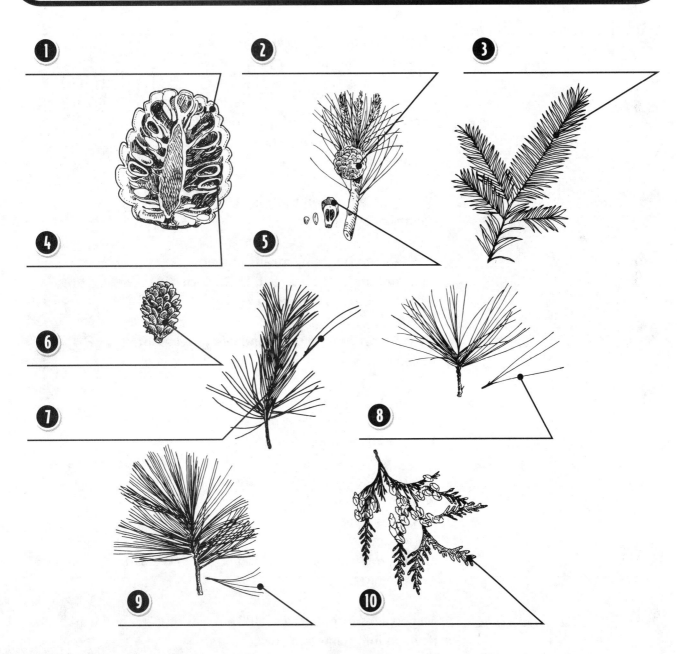

Name _____ Date _____

Life Cycle of a Conifer

Conifers are woody trees and shrubs with needlelike or scalelike leaves. Most are evergreen. Conifers include redwoods, junipers, spruces, pines, and firs. The reproductive cycle of a conifer uses cones. Match each term in the word box to its description.

> pollen grains   male cones   cone   fertilization   embryo
> female cones   gametophyte   seedling   pollination   wind currents

1 _____ This is a cluster of modified leaves where the spore-producing parts of the plant develop.

2 _____ This is the phase of a plant's life cycle in which either sperm or eggs are produced.

3 _____ Every spring, millions of these drift off of male cones and some land on female cones.

4 _____ This is the term used for the young conifer plant.

5 _____ Conifers reproduce by means of seeds that contain food tissue and this, which will grow into a plant.

6 _____ Microspores form in these and will develop into pollen grains.

7 _____ Conifer pollination is dependent on these to blow the yellow pollen from the male cones to the female cones.

8 _____ This describes the joining of the pollen grain with an egg cell and can occur anywhere from a month to a year after pollination.

9 _____ This describes the arrival of the pollen grains onto the ovules of female cones.

10 _____ Megaspores form in the shelflike scales of these and will develop into gametophytes.

The Angiosperms

Angiosperms are flowering, seed-bearing plants. Most of the plants we see every day are angiosperms. Some have flowers so small we do not know they are flowers, such as in grass or on an oak tree. Others have flowers in a wide range of colors, sizes, and shapes. Use the terms in the word box to label the diagram.

flower	fruit	seeds
leaf	stem	root

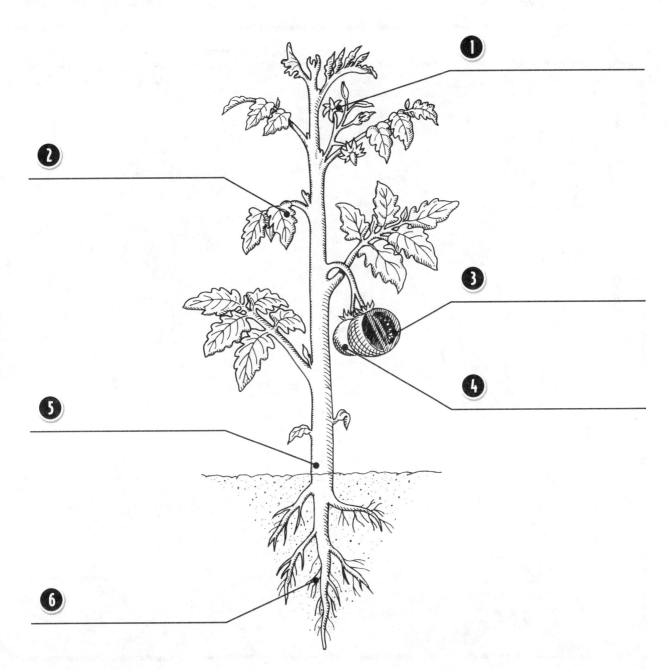

Name _____   Date _____

Monocots and Dicots

The angiosperms can be divided into two groups: the monocots and the dicots. *Monocot* refers to the fact that these plants have seeds with only one cotyledon or seed coat. Corn, wheat, and rice are monocots. *Dicot* refers to two cotyledons. Beans, roses, and dandelions are examples of dicots. Use the phrases in the word box to classify the characteristics of monocots and dicots.

one seed leaf	two seed leaves	parallel leaf veins
netlike leaf veins	flower parts in threes	scattered bundles
flower parts in fours or fives	bundles in a ring	branching fibrous roots
central taproot		

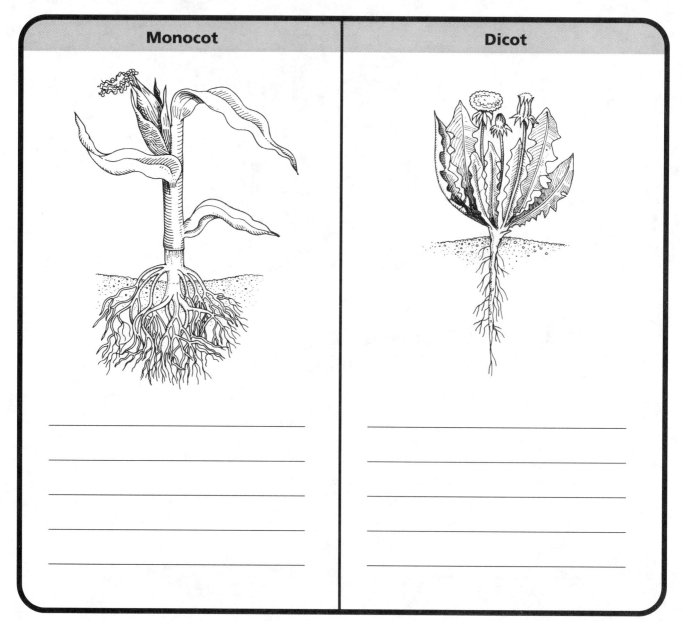

Monocot

_____

_____

_____

_____

Dicot

_____

_____

_____

_____

Name _____ Date _____

Flowering Plant Parts

All angiosperms have the same basic parts, even though these parts might look very different. The parts of a flowering plant assist in pollination, the process by which pollen and the eggs of plants are joined to produce new seeds. Match each term in the word box to its definition.

petal	sepal	bud
leaf	stem	root
ovule	fruit	seed
pollen	pollinator	calyx

1 _____ This is an air current, water current, insect, bird, or animal that makes the transfer of pollen.

2 _____ This is a structure within the ovary that holds a plant egg cell. After fertilization, it will become a seed.

3 _____ This is the name given to the expanded and ripened ovary of a plant.

4 _____ This is one of the often brightly colored parts of a flower immediately surrounding the reproductive organs.

5 _____ This refers to an undeveloped plant part, either a leaf or a flower.

6 _____ These are the fine powderlike grains that contain sperm-bearing cells of a plant.

7 _____ This plant structure contains chlorophyll and is the major region of photosynthesis.

8 _____ This is a fully mature ovule which contains the plant embryo and supporting parts.

9 _____ This descending part of a plant anchors the above ground parts and usually stores food.

10 _____ This is one of the usually green parts formed at the base of a petal.

11 _____ This is another name for the stalk that supports the various aboveground parts of a plant.

12 _____ This is the name for all the sepals of a flower.

Root Systems

Two types of root systems are found in flowering plants. Use the terms in the word box to label the diagrams.

fibrous roots	root hair cell	taproot system
lateral root	prop roots	crown
stems and leaves	root tip	

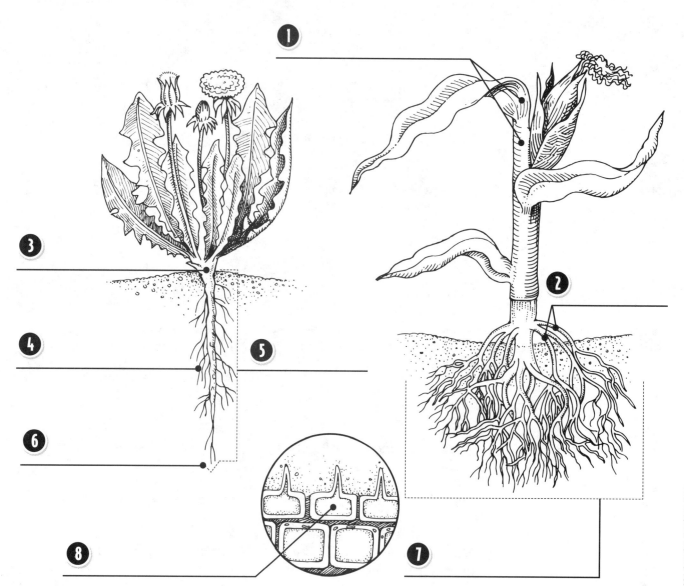

Inside a Root

Roots are specialized structures that descend into the soil as they spread downward and outward. A plant's root system absorbs water and dissolved nutrients to store as food, releasing it as the plant needs it. The roots also serve to anchor the rest of the plant. Use the terms in the word box to label the diagrams.

root hair	epidermis	lateral root
root cortex	root tip	vascular cylinder
primary xylem	primary phloem	root cap

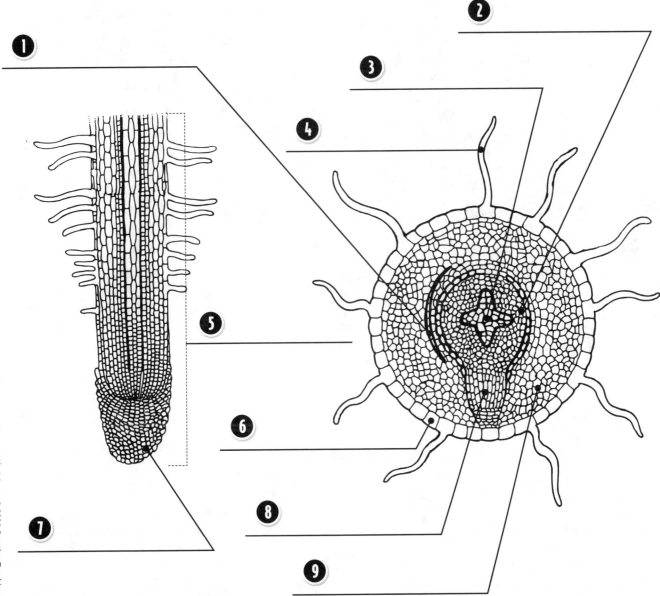

Life Science © 2004 Creative Teaching Press

Underground Stems

Tubers, rhizomes, and bulbs are three types of underground stems. Use the terms in the word box to label each type of stem base and its parts. Some terms are used more than once.

| bud | stem | leaf | root | bulb | rhizome | tuber |

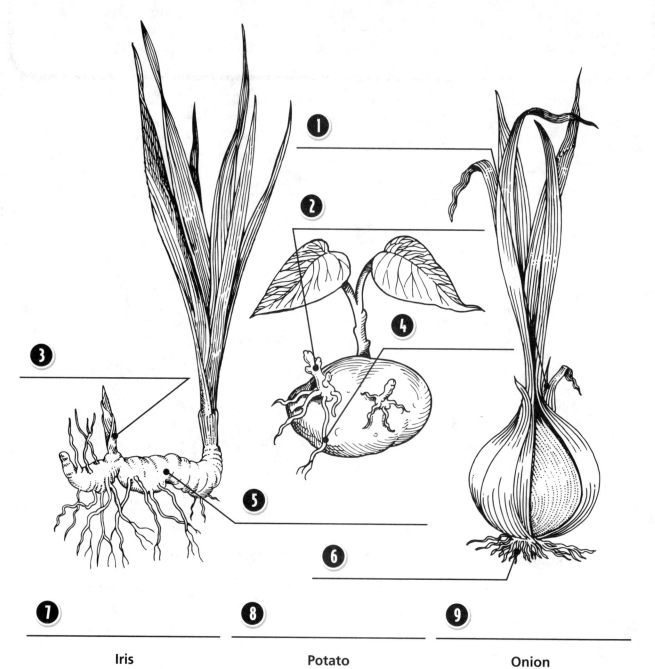

① _____

② _____

③ _____

④

⑤

⑥ _____

⑦ _____
Iris

⑧ _____
Potato

⑨ _____
Onion

Stems Above Ground

Stem structures provide support for the plant and its leaves and flowers. Stems transport water and nutrients from the roots and throughout the plant parts. There are differences between the stems of monocots and dicots. Use the terms in the word box to label the diagrams. Some terms are used more than once.

vascular bundles	pith	cortex
node	ground tissue	petiole
bud	blade	sheath

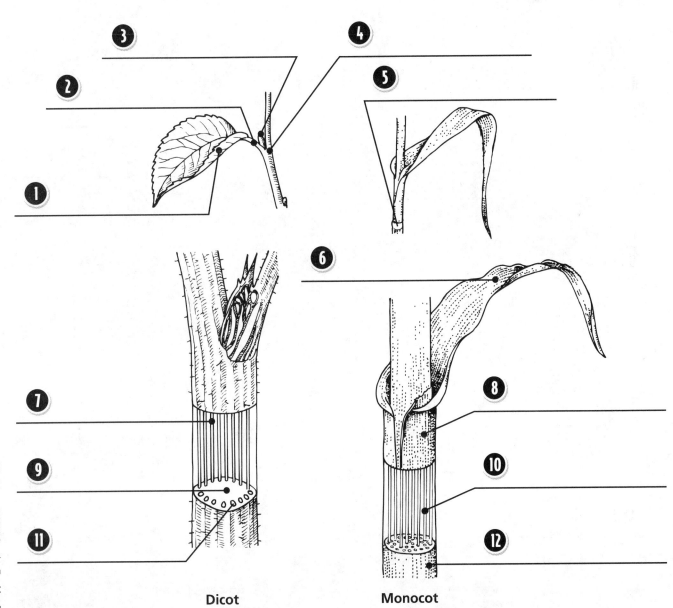

Dicot　　　　**Monocot**

Woody or Herbaceous Stems

The stems of woody plants are different than those of herbaceous plants. Herbaceous plants are those with soft or fleshy tissue. These plants often die back to the ground each year. Woody plants have hardened woody stems. These plants do not die back, instead adding a new layer of growth to their stem each year. Use the terms in the word box to label the diagrams. Some terms are used more than once.

| cork | bark | pith | cortex |
| cambium | phloem | xylem | epidermis |

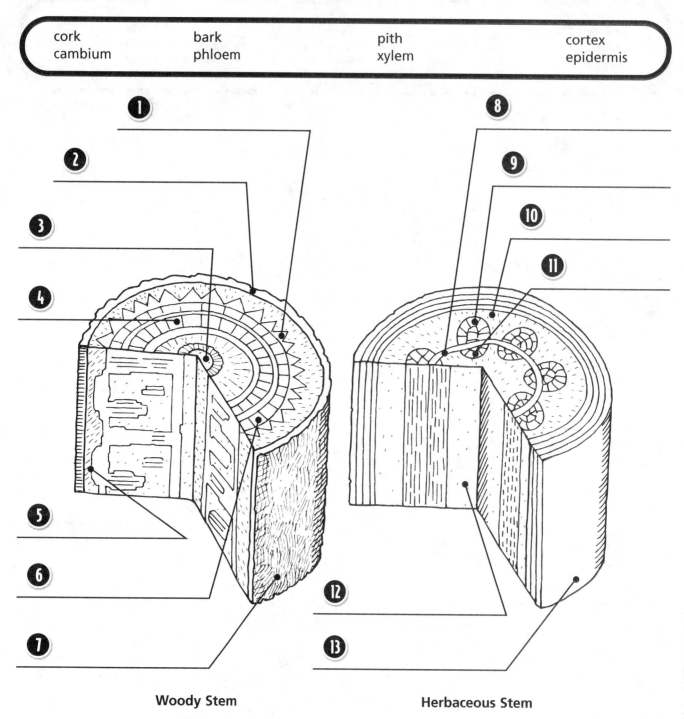

Woody Stem   **Herbaceous Stem**

Name _____ Date _____

Really Big Stems

The trunks of trees are older woody stems with massive secondary growth. Each year the tree is alive, it adds another layer to its stem or trunk. Each of these layers is referred to as a growth ring. Use the terms in the word box to label the diagram.

leaves	bark	trunk
phloem	sapwood	roots
cambium	heartwood	branches
xylem		

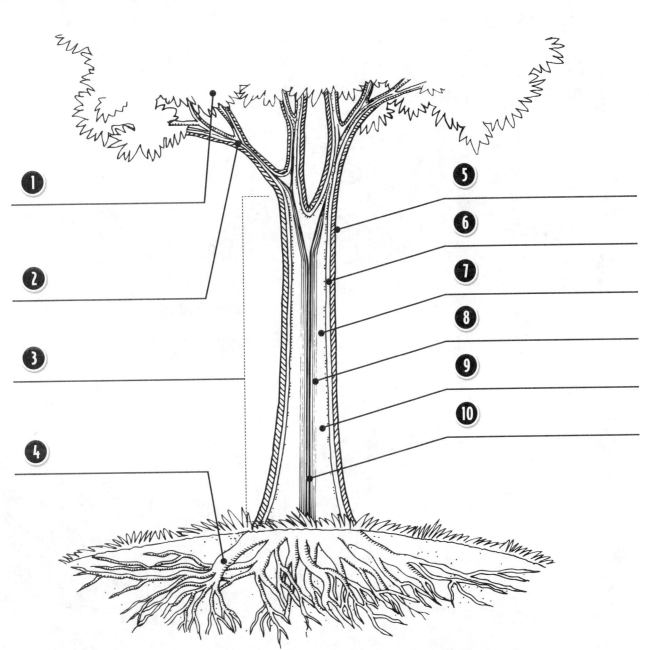

A Look on the Inside

By looking at the cross section of a tree, we can see the specialized tissues that make up its trunk. Use the terms in the word box to label the diagram. Some terms are used more than once.

> heartwood   phloem   sapwood
> vascular cambium   bark   xylem

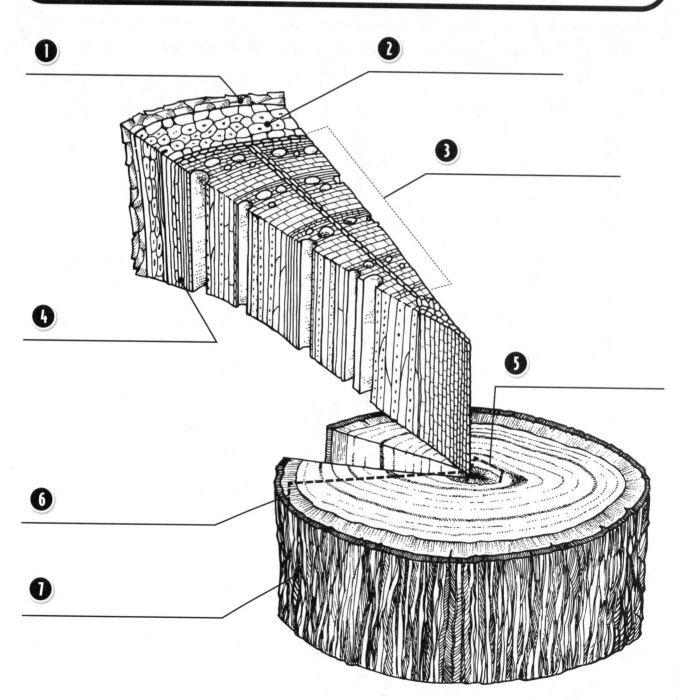

Life Science © 2004 Creative Teaching Press

Name _____ Date _____

As a Tree Grows

While the trunk of a tree may be massive, it is the tips of its branches that keep on growing up and out. Use the terms in the word box to label the diagram.

terminal bud	leaf bud	flower bud
leaf scar	node	internode
bud-scale scar	branch	

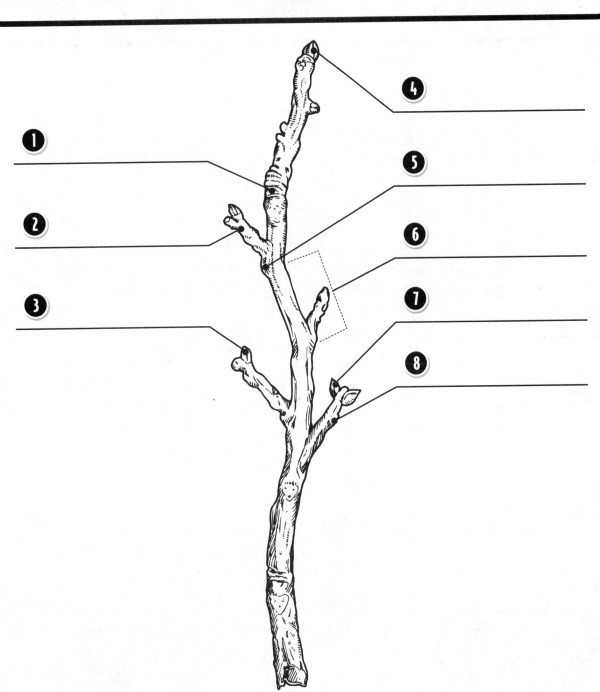

1 _____

2 _____

3 _____

4 _____

5 _____

6 _____

7 _____

8 _____

Life Science © 2004 Creative Teaching Press

Looking at Leaves

The leaves of various angiosperms have characteristics that make them distinguishable from one another. In fact, plants are classified by the type of leaves they have. Use the terms in the word box to label the diagrams.

petiole	blade	leaf margin
leaflet	veins	lobe
simple leaf	compound leaf	

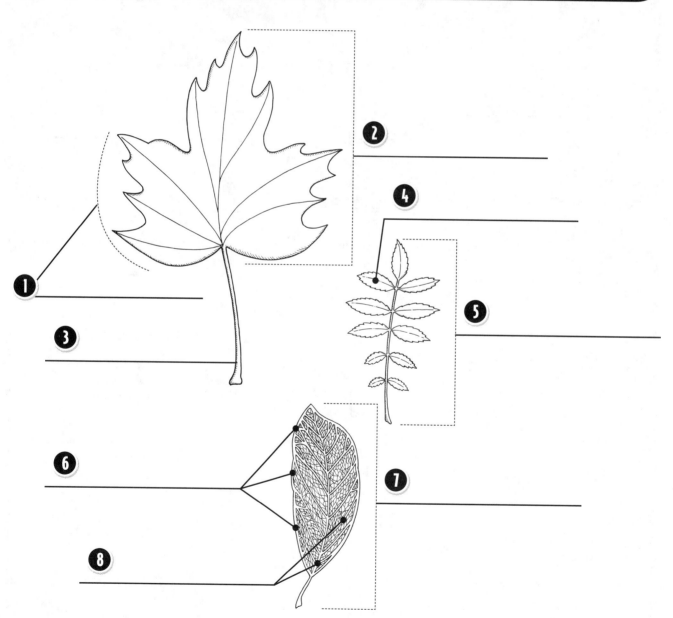

Life Science © 2004 Creative Teaching Press

Characteristics of Leaves, Part One

Use the terms in the word box to complete the chart.

simple	palmate	pinnate
v-shaped	rounded	flat
heart	uneven	compound

Division of Leaf

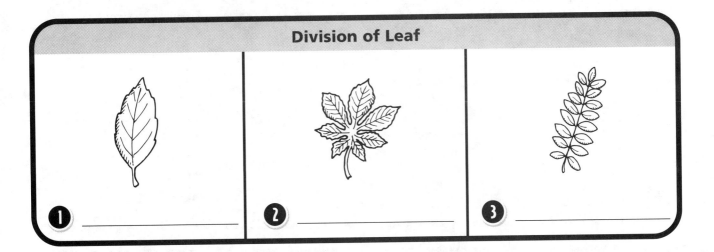

1 _____

2 _____

3 _____

Overall Shape of Leaf Base

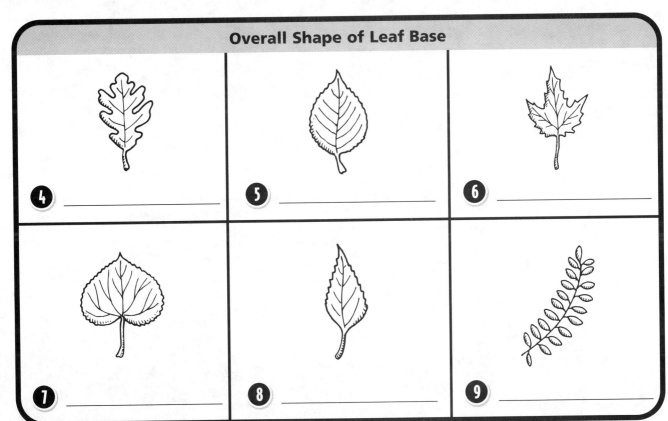

4 _____

5 _____

6 _____

7 _____

8 _____

9 _____

Name _____ Date _____

Characteristics of Leaves, Part Two

Use the terms in the word box to complete the chart.

parallel	palmate	pinnate
lobed	double saw-toothed	saw-toothed
smooth	opposite	alternate

Type of Leaf Margin

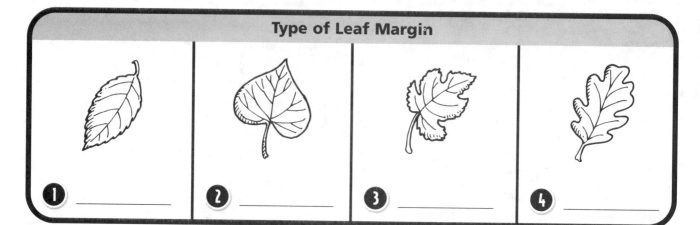

1 _____   2 _____   3 _____   4 _____

Type of Vein Pattern

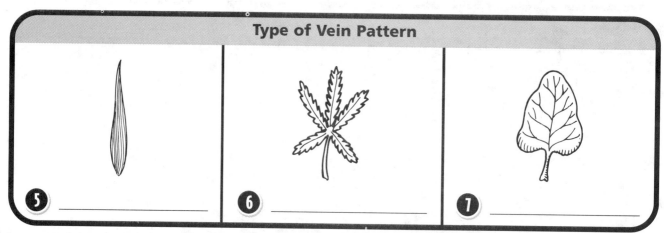

5 _____   6 _____   7 _____

How Leaves Are Arranged on the Stem

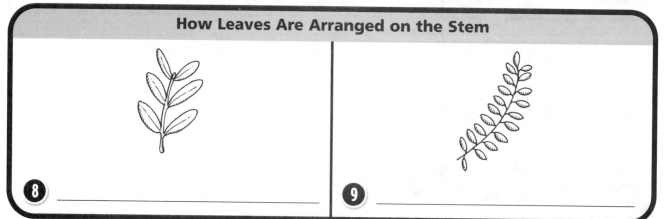

8 _____   9 _____

A Closer Look

Leaves are the food factories for green plants. It is the presence of chlorophyll in leaves that allows the plants to manufacture their own food. Use the terms in the word box to label the diagrams. Some terms are used more than once.

guard cells	waxy layer	stomata
vein	epidermis	spongy layer
palisade layer		

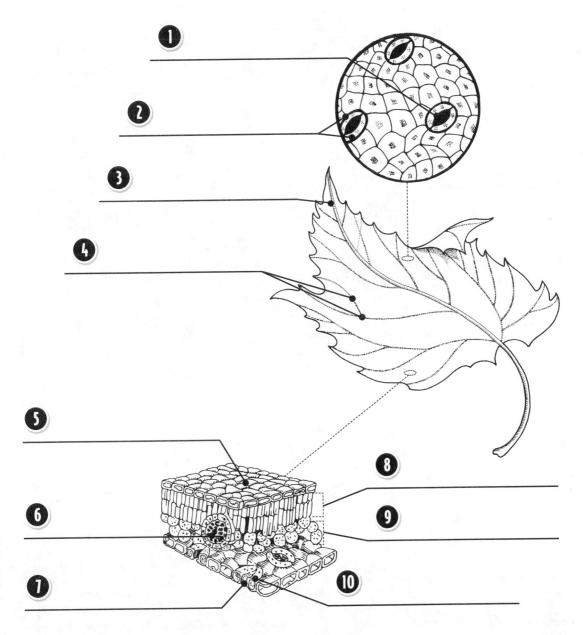

Seed-Producing Parts of a Flower

As pretty as flowers can be, their main function is to attract pollinators in order to produce seeds. Use the terms in the word box to label the diagram.

stamen	receptacle	stigma
ovary	carpel	anther
style	filament	ovule

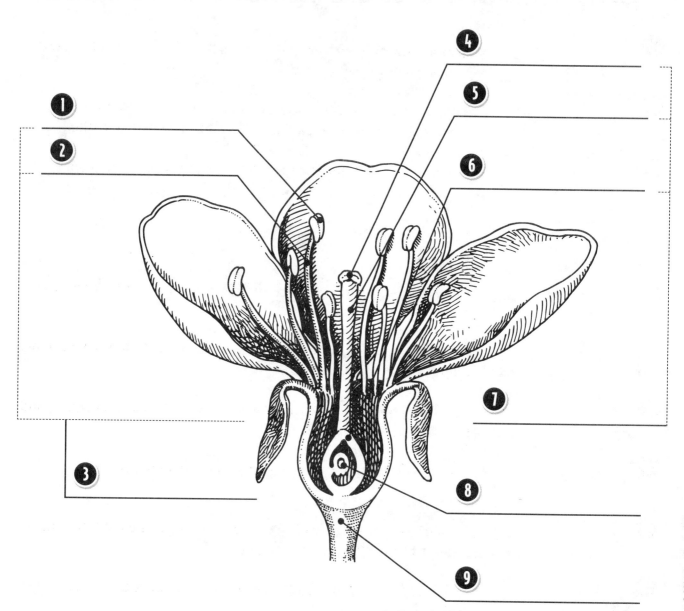

Name _____ Date _____

The Process of Pollination

Pollination is the way pollen grains are transferred from the male parts to the female parts of a plant. Grains, fruits, vegetables, flowers, and trees must be pollinated in order to produce seeds or fruit. Match each term in the word box to its definition.

stamens	corolla	pollen grains
ovary	carpel	anther
style	filament	stigma
perfect		

1. _____ All of a flower's petals combined form this structure. Often its shape and color attract animals that serve as pollinators.

2. _____ These are the male reproductive parts of a flower and are found just within a flower's corolla. Each includes a filament and an anther.

3. _____ This structure holds the chambers in which the pollen is formed.

4. _____ This is the stalk that holds the anther.

5. _____ These sperm-producing bodies are transferred in order to fertilize a plant.

6. _____ This is the female reproductive part of a plant and includes the stigma, the style, and the ovary.

7. _____ This is the chamber in which the eggs develop, fertilization occurs, and the seeds develop.

8. _____ This is a sticky or hairy stalk that captures pollen grains.

9. _____ A slender extension of the ovary's walls, this structure holds the stigma up high in the path of pollen grains.

10. _____ This is a term used to describe a flower that contains both male and female parts.

Name _____ Date _____

Monocots

Monocots are plants that have only one cotelydon or seed leaf. A corn plant is an example of a monocot. Use the terms in the word box to label the diagram. Some terms are used more than once

cotyledon   endosperm   embryo
seed and fruit coats   radicle   epicotyl
hypocotyl

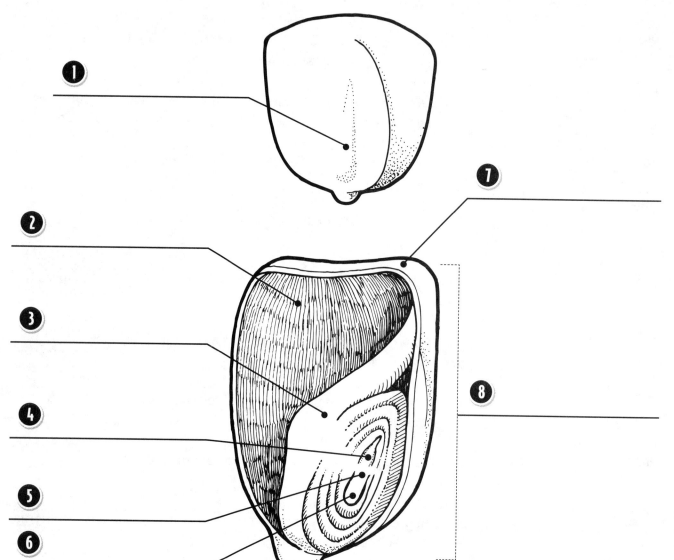

Name _____ Date _____

Dicots

Dicots are plants with two cotyledons or seed leaves. An example of a dicot is a bean seed. Use the terms in the word box to label the diagram.

cotyledons	seed coat	embryo
hilum	hypocotyl	radicle
epicotyl		

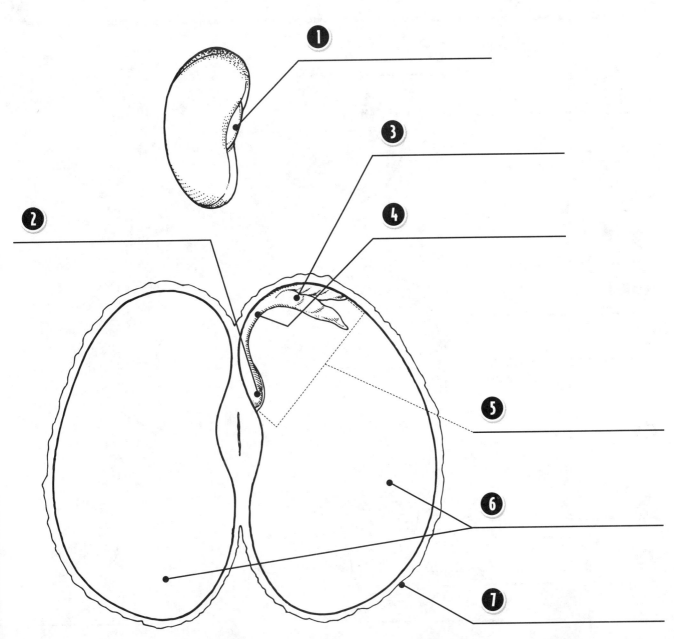

Growth of a Monocot Plant

Following germination, the seed coat splits and the primary root begins to grow first. Then the plant pushes up out of the soil, reaching toward the light. As it grows, the young plant uses the stored energy in the cotyledon until its first leaves emerge and photosynthesis can begin. Use the terms in the word box to label the diagram.

first foliage leaf	seed coat	primary root
leaf sheath	prop root	fibrous roots
first internode	primary leaf	branch root

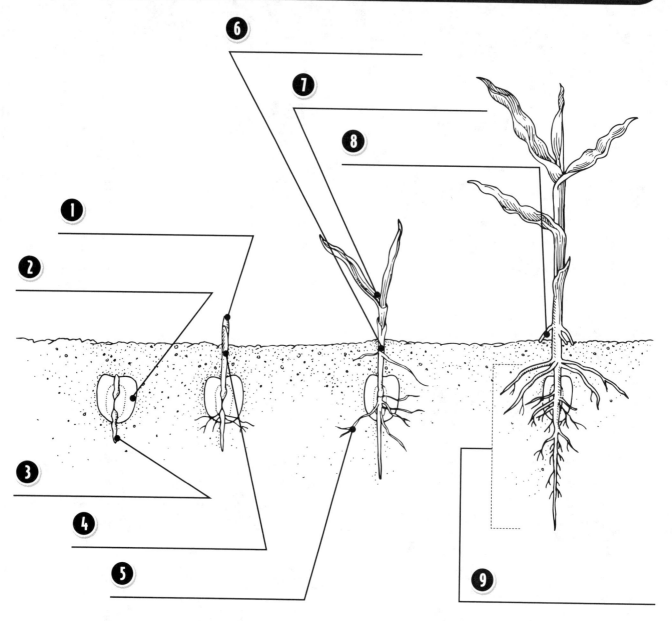

Name _____ Date _____

Growth of a Dicot Plant

A bean plant begins to grow by sending down its primary roots. Then the two cotyledons burst through the soil as the stem begins to grow. Once the primary leaves are formed, the cotyledons shrivel up and fall away. Use the terms in the word box to label the diagram.

foliage leaf
node
withered cotyledons
primary leaf

point where cotyledons were attached
branch roots
seed coat   primary root
hypocotyl   cotyledons

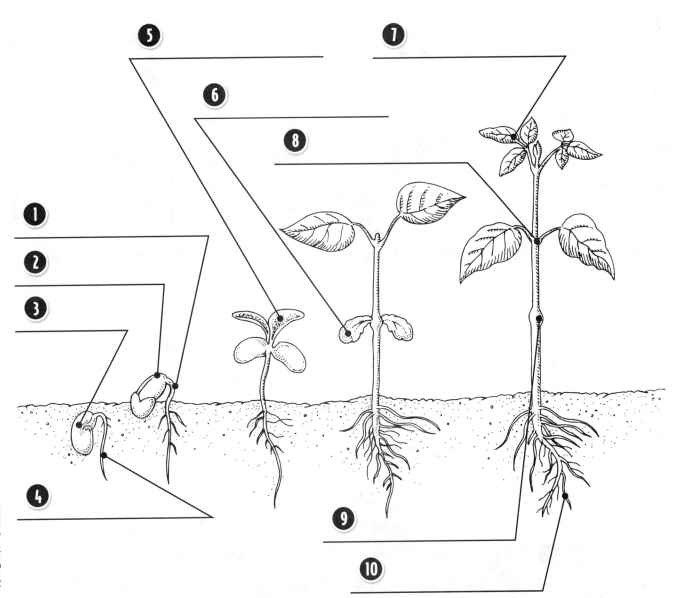

Members of the Animal Kingdom

Kingdom Animalia is divided into subgroups called phyla. Each phylum displays certain characteristics. Use the definitions below to identify each animal representative of some of the animal phyla.

Porifera: This phylum includes the sponges. They have no symmetry, tissues, or organs.

Cnidaria: This phylum includes jellyfish, coral, and sea anemones. They have radial symmetry and stinging cells.

Platyhelminthes: This phylum includes flatworms. They have flattened bodies and bilateral symmetry.

Mollusca: The mollusks includes soft-bodied animals with and without shells such as clams, snails, slugs, and squids.

Annelida: This phylum includes worms that have segmented bodies.

Arthropoda: The most successful phylum on Earth, arthropods include crustaceans, spiders, and insects.

Echinodermata: The echinoderms are spiny-skinned animals such as the sea star and sea urchin.

Chordata: The chordates have bilateral symmetry and a spinal nerve cord. Most of this phylum is made up of the vertebrates, or animals with backbones.

Name _____ Date _____

A Typical Animal Cell

While the cells found in various tissues of animals are unique and specialized, the basic structure of animal cells is the same. Animal cells lack the rigid cell wall found in plant cells. Use the terms in the word box to label the diagram.

cytoplasm   nucleolus   lysosome
cell membrane   nucleus   ribosomes
endoplasmic reticulum   mitochondrion   vacuole

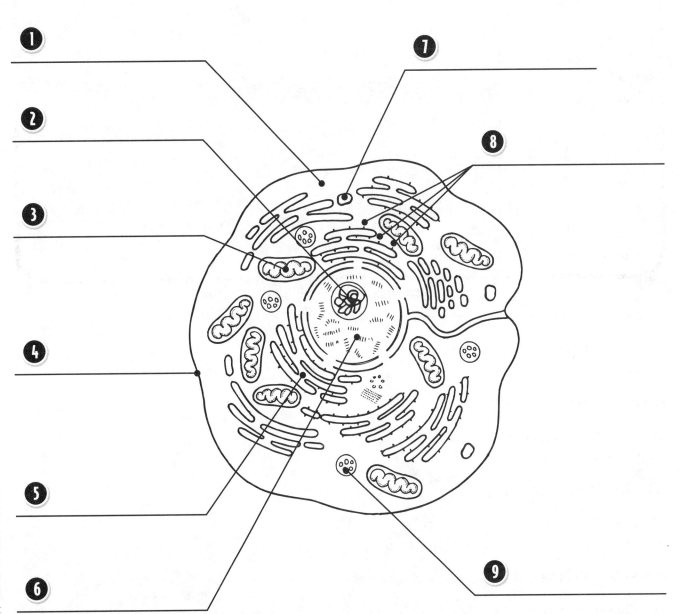

Functions within an Animal Cell

The parts within a cell are called organelles. Each organelle has a unique function that it serves for the cell. Match each term in the word box to its definition.

cytoplasm   nucleolus   lysosome
cell membrane   nucleus   endoplasmic reticulum
mitochondria   vacuole

1 _____ Located in the nucleus, this organelle is made up of RNA and protein.

2 _____ These structures are passageways from the nucleus that transport proteins through the cell.

3 _____ This organelle is the control center of the cell. It contains chromosomes and DNA.

4 _____ This substance contains all the living material in the cell.

5 _____ These organelles break down glucose to supply the cell with energy.

6 _____ This small round organelle is involved in digestion.

7 _____ This expanding and contracting organelle stores water, nutrients, and wastes.

8 _____ This enclosing structure holds the cell together and controls what moves into and out of the cell.

Name _____ Date _____

Body Symmetry

Within the animal kingdom, bodies can be described by their symmetry. The symmetry of an organism affects how tissues and organs are arranged within the body. Use the definitions below to label the symmetry of each animal.

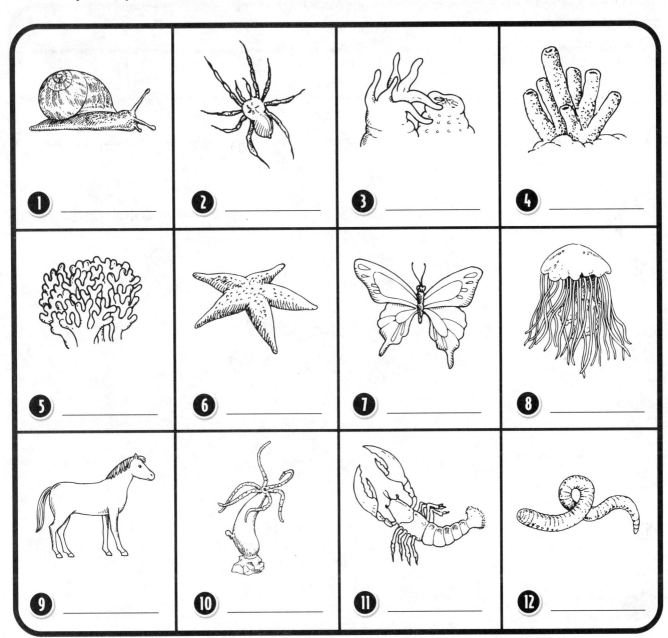

① _____

② _____

③ _____

④ _____

⑤ _____

⑥ _____

⑦ _____

⑧ _____

⑨ _____

⑩ _____

⑪ _____

⑫ _____

Radial: Body parts are symmetrical around a central point. Body parts radiate out from this central point.

Bilateral: The left and right sides of a body are alike and of equal proportion.

Asymmetrical: The body has no definite shape and there is no symmetry.

The Sponges

Sponges are invertebrate animals that live mostly in marine environments. In order to feed, they bring in water through the ostia (plural for ostium), collect the minute particles, and move the water out through the uppermost opening, the osculum. The collar cells both help move the water through the sponge and collect the tiny nutrients it needs for food. Use the terms in the word box to label the diagram.

osculum	spicules	collar cell
epithelial cell	ostium	flagellum
nucleus	water flow	

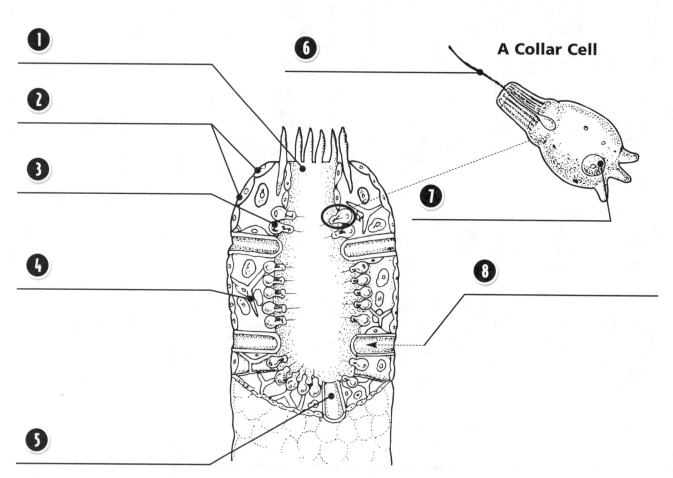

A Collar Cell

① ② ③ ④ ⑤ ⑥ ⑦ ⑧

Name _____   Date _____

The Hydra

A hydra is related to sea anemones, jellyfish, and coral. It is a simple, hollow animal with a foot at one end and an opening at the other that forms the mouth. The six to ten tentacles that surround the mouth are used for capturing food. It moves either by gliding on the foot, somewhat like a snail, or by somersaulting. Use the terms in the word box to label the diagram.

tentacle	mouth	bud
ovary	nematocyst	foot
ectoderm	endoderm	mesoglea
gastrovascular cavity		

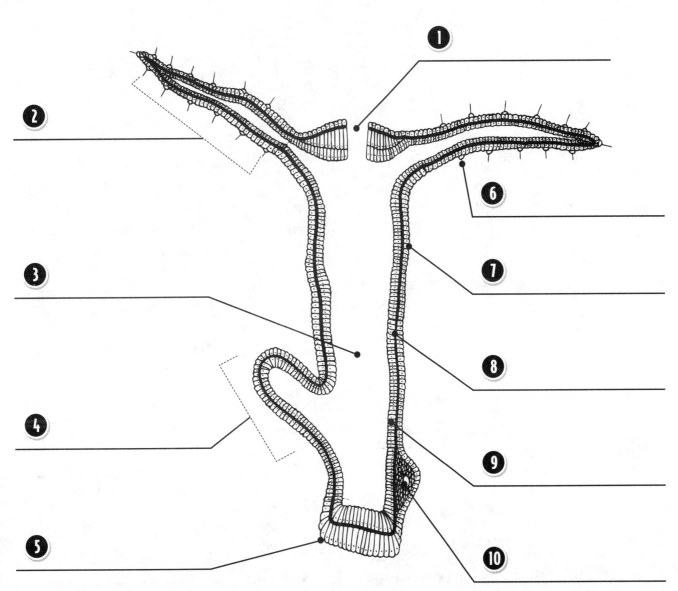

Name _____ Date _____

Flatworms

Flatworms are soft-bodied and are the simplest of animals that have heads. They have bilateral symmetry and flattened bodies. Three main classes of flatworm are tapeworms, flukes, and planarians. Tapeworms and flukes are parasites. The planarian is freeliving. Use the terms in the word box to label the diagrams of a planarian.

pharynx	branching gut	eyespot
brain	ovary	testis
protonephridia	nerve cord	

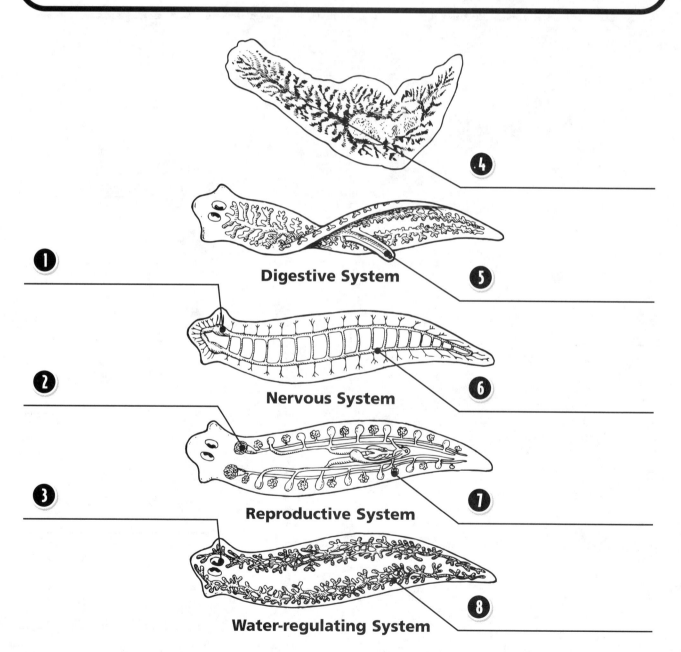

Digestive System

Nervous System

Reproductive System

Water-regulating System

Life Science © 2004 Creative Teaching Press

Name _____ Date _____

Mollusks—A Snail

The mollusks are soft-bodied animals. All mollusks are bilateral and have a mantle, a tissue fold that hangs like a skirt over the body. Many mollusks have shells. The three groups of mollusks are the gastropods, the bivalves, and the cephalopods. Gastropods include snails and slugs. Use the terms in the word box to label the diagram.

gill	anus	mantle cavity
radula	excretory organ	heart
digestive gland	stomach	mantle
shell	foot	mouth

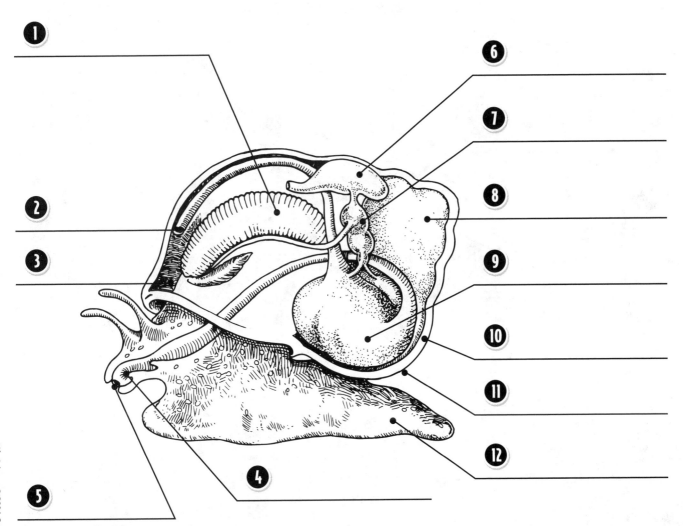

Mollusks—A Clam

A clam is an example of a bivalve mollusk. Bivalve mollusks have two shells that enclose the body. A bivalve siphons water through the mantle cavity where mucus on the gills traps bits of food. Then the remaining water is siphoned back out of the body. Use the terms in the word box to label the diagram. Some terms are used more than once.

mouth	muscle	palps
exhalant siphon	inhalant siphon	shell
mantle	gill	foot

A Clam with One Shell Removed

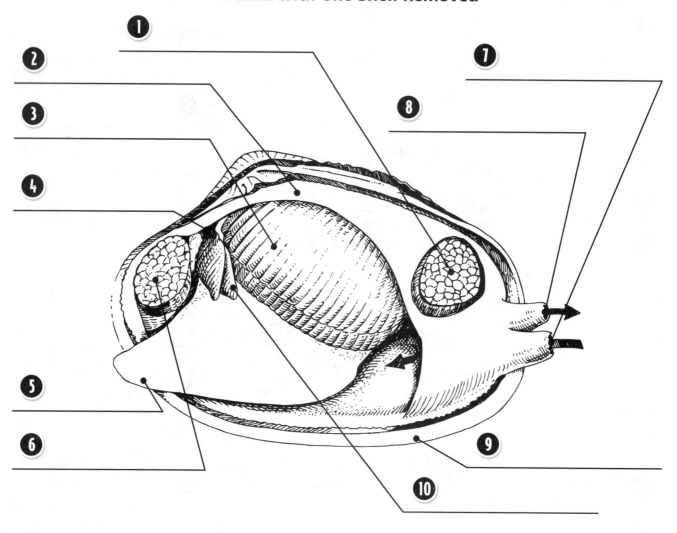

Mollusks—A Squid

Cephalopods are mollusks that include octopuses, cuttlefish, and squids. The foot of these mollusks is divided into arms and tentacles around the mouth, which are used to capture prey. Cephalopods are able to move through their environment with jet propulsion, drawing water into and out of the mantle cavity through a siphon. Use the terms in the word box to label the diagram.

mantle	internal shell	stomach
kidney	digestive gland	radula
arm	jaw	tentacle
siphon	esophagus	brain
ink sac	accessory heart	heart
gill	reproductive organ	anus

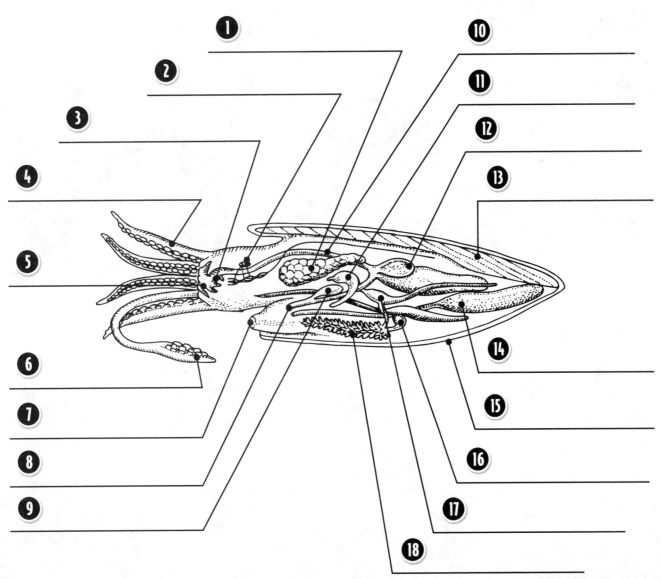

Segmented Worms

The segmented worms include earthworms and leeches. Both of these animals display the body segmentation characteristic of this class. Most of the circulatory, nervous, and digestive systems are located near the head. The rest of the segments are mostly identical to each other. Earthworms are hermaphroditic, which means each one contains both male and female reproductive organs.

setae	mouth	gizzard
intestine	hearts	dorsal blood vessel
clitellum	ventral blood vessel	crop
anus	esophagus	

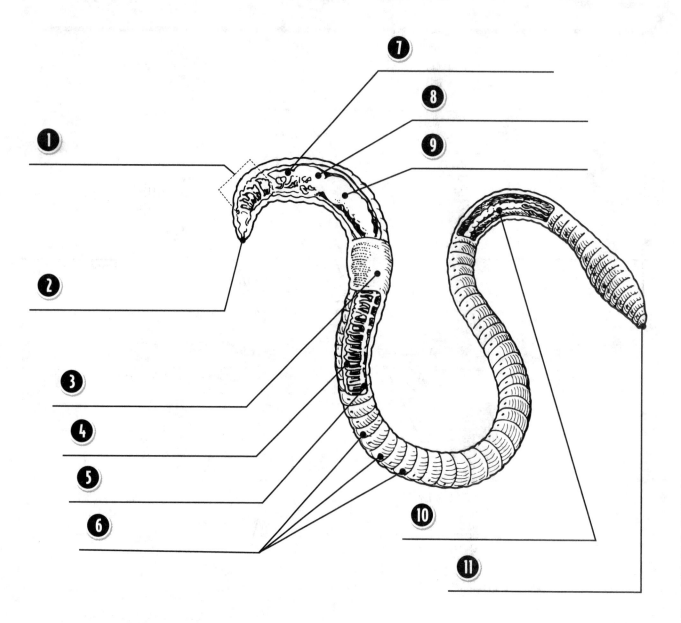

The Arthropods

The **arthropods** are animals that have jointed legs. This group of animals is the most successful on Earth because of the number of species, the number of offspring produced, and how well they are able to adapt to their environment. Use the terms in the word box to complete the chart. Some terms are used more than once.

rounded	segmented body	flat
two pairs of legs per segment	flexible exoskeleton	gills
one pair of legs per segment	hard	no antennae
two pairs of antennae	four pairs of legs	three pairs of legs
two body sections	three body sections	one pair of antennae

Diplopoda

_____

_____

Chilopoda

_____

_____

The Crustaceans

_____

_____

_____

The Arachnids

_____

_____

_____

The Insects

_____

_____

_____

Name _____ Date _____

A Crustacean—The Crayfish

Crustaceans get their name from their hard yet flexible "crust" or exoskeleton. Their bodies have two parts, they breathe with gills, and they have two pairs of antennae. Shrimps, lobsters, crabs, barnacles, and pillbugs are all examples of crustaceans. Use the terms in the word box to label the diagram of a crayfish.

antennae	cephalothorax	abdomen
maxillipeds	cheliped	walking legs
swimmerets	tailfan	eye

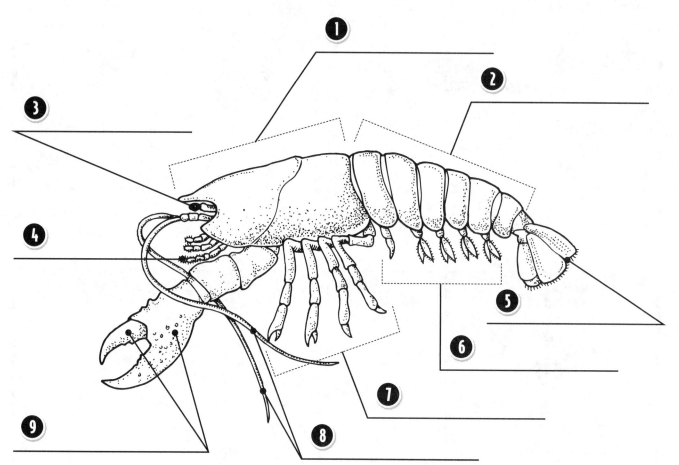

A Close Look at Arachnids

Arachnids include spiders, ticks, scorpions, and horseshoe crabs. While related to insects, they are not insects. Arachnids have four pairs of legs and a two-part body. Use the terms in the word box to label the diagram of a female web-spinning spider.

poison gland	eye	heart
anus	spinnerets	silk gland
ovary	book lung	mouth
chelicera	pedipalp	cephalothorax
abdomen		

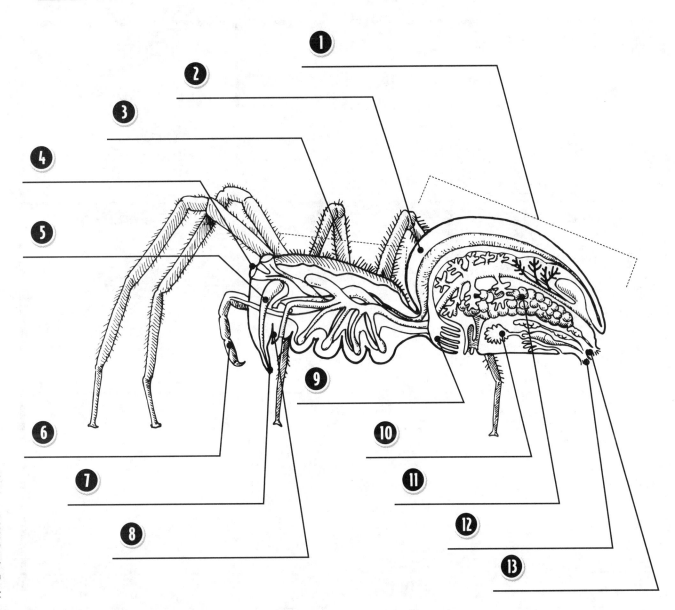

In the Right Order

Insects are invertebrates with jointed legs, segmented bodies, and an exoskeleton. There are more than twenty-five different orders of insects. These are seven of the more common orders. Use the terms in the word box to complete the chart.

stag beetle	butterfly	honeybee
ladybug	grasshopper	mosquito
ant	praying mantis	back swimmer
giant water bug	silkworm moth	dragonfly
damselfly	housefly	

Coleoptera	**Diptera**	**Hemiptera**
_____	_____	_____
_____	_____	_____

Lepidoptera	**Hymenoptera**	**Odonata**
_____	_____	_____
_____	_____	_____

Orthoptera
_____
_____

The Honeybee

The honeybee is the common name for any of the highly social bees. Honeybees are known for their ability to make and store honey and their ability to be domesticated. The honeybee provides pollination for many crops and plants. Honeybees have a complex social structure. Each hive has one queen and many drones and worker bees. They are among the most studied and best-known insects. Use the terms in the word box to label the diagram.

simple eyes	antennae	abdomen	thorax	head
ganglia	stinger apparatus	intestine	pollen basket	mandible
compound eye	heart	rectum	honey stomach	stinger

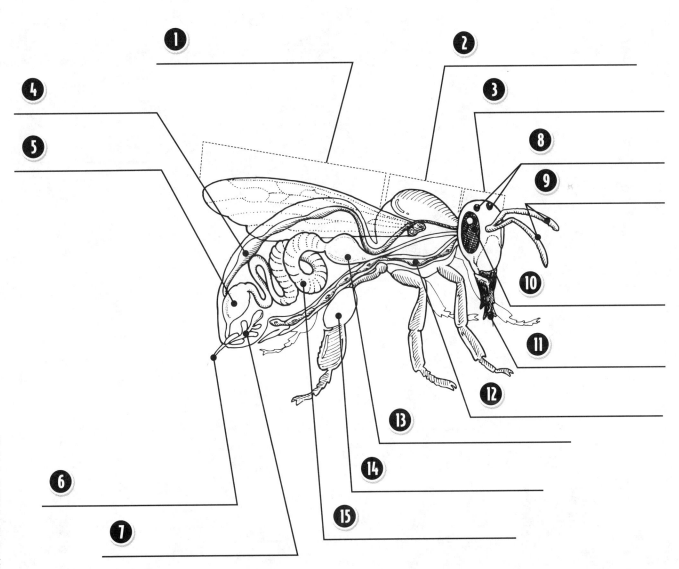

Life Cycle of a Bee

The queen is the only sexually productive female in the honeybee colony. She is the mother of all drones, workers, and future queens. She can lay as many as 1,500 eggs in a day. These eggs are then attended by the worker bees. Once the eggs are laid, they go through distinct stages of development. Use the phrases in the word box to label each stage in the diagram.

grub sealed in its cell	grub fed by worker	egg laid by queen
full-grown bee grub	young adult leaves cell	grub becomes a pupa

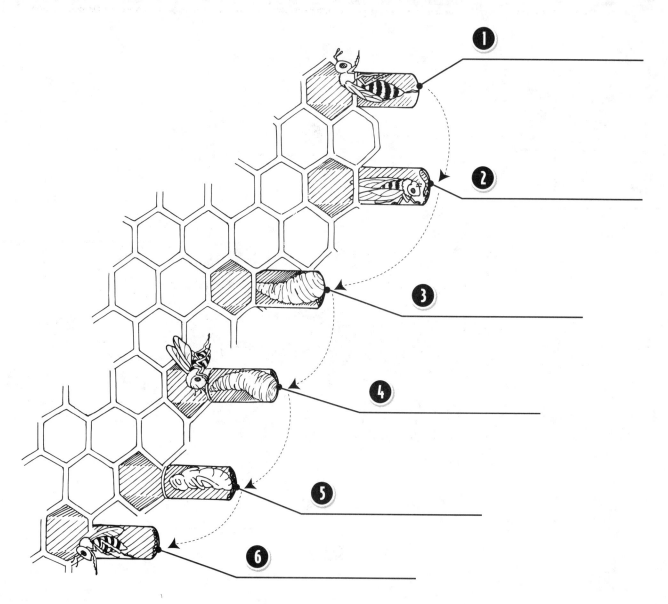

Name _____ Date _____

A Grasshopper

Grasshopper is the common name for any of the winged insects with hind legs adapted for hopping. Only the adults can fly. Male grasshoppers can make chirping sounds by rubbing their hind legs against other parts of their bodies. The spiracles are tiny circular openings grasshoppers use to obtain oxygen. Use the terms in the word box to label the diagram.

femur	head	knee joint
antennae	jumping leg	abdomen
thorax	walking legs	ovipositor
front wing	hind wing	compound eye
spiracle		

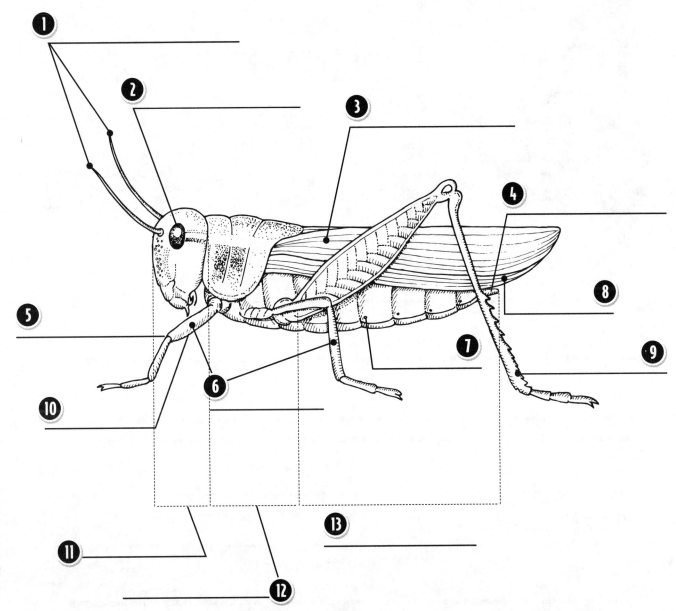

Name _____ Date _____

Butterflies and Moths

Butterflies and moths are insects with four wings, each covered with tiny, shingle-like scales. Most butterflies and moths use distinctive mouthparts to feed on the nectar of certain flowers. However, butterflies and moths each have characteristics that can be used to tell them apart. Use the terms in the word box to label the diagrams.

forewing	compound eye	hind wing
legs	proboscis	antennae
abdomen		

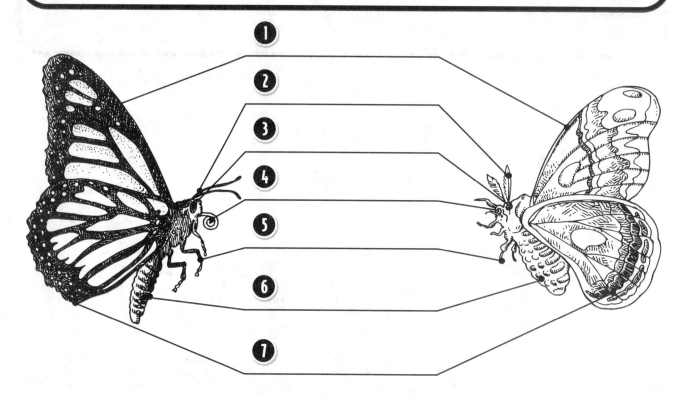

Use the terms in the word box to complete the chart.

wings held upright or flat	wings folded	chrysalis
cocoon	smooth with end knobs	feathered with no knobs

	Resting	**Antennae**	**Development**
Butterfly			
Moth			

Metamorphosis

Insects go through a process called metamorphosis during their life cycle. In complete metamorphosis, there are four distinct stages. In incomplete metamorphosis, the young resembles the adult. As it grows, the animal gradually changes through molting, or shedding. Use the terms in the word box to label each diagram. Some terms are used more than once.

Complete	Incomplete	adult
egg	larva	pupa
nymph		

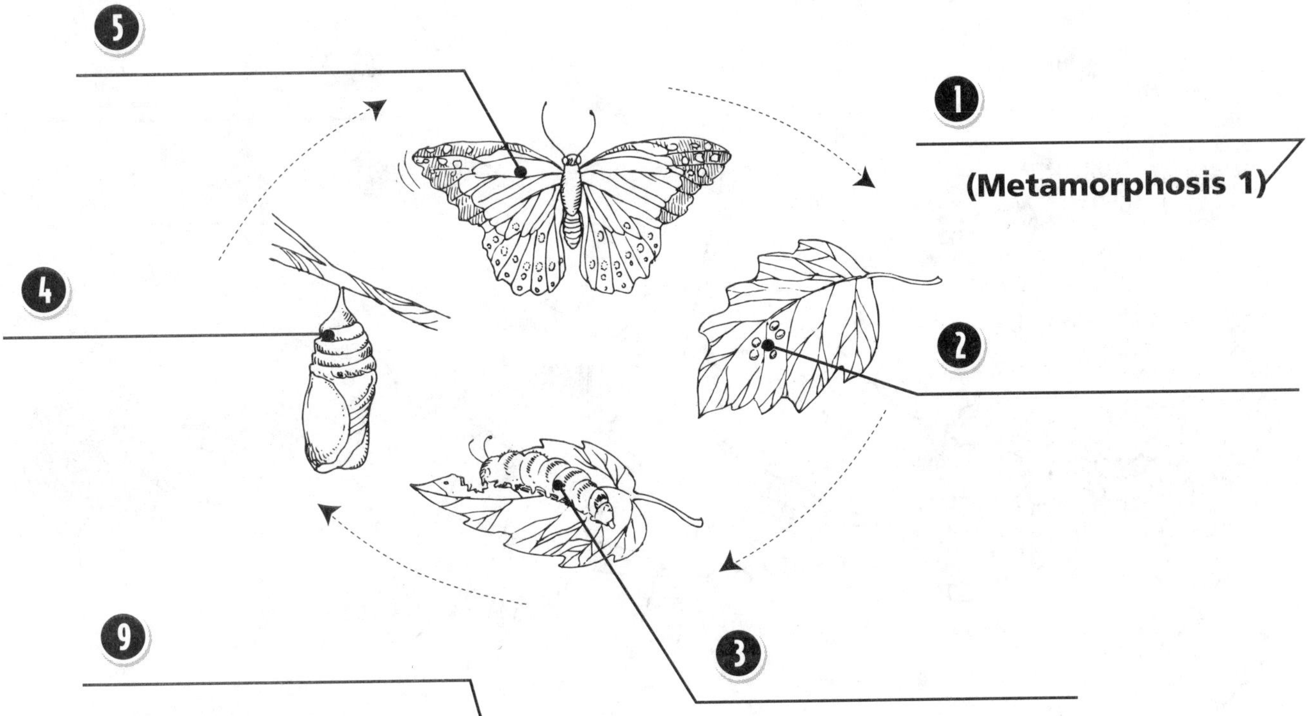

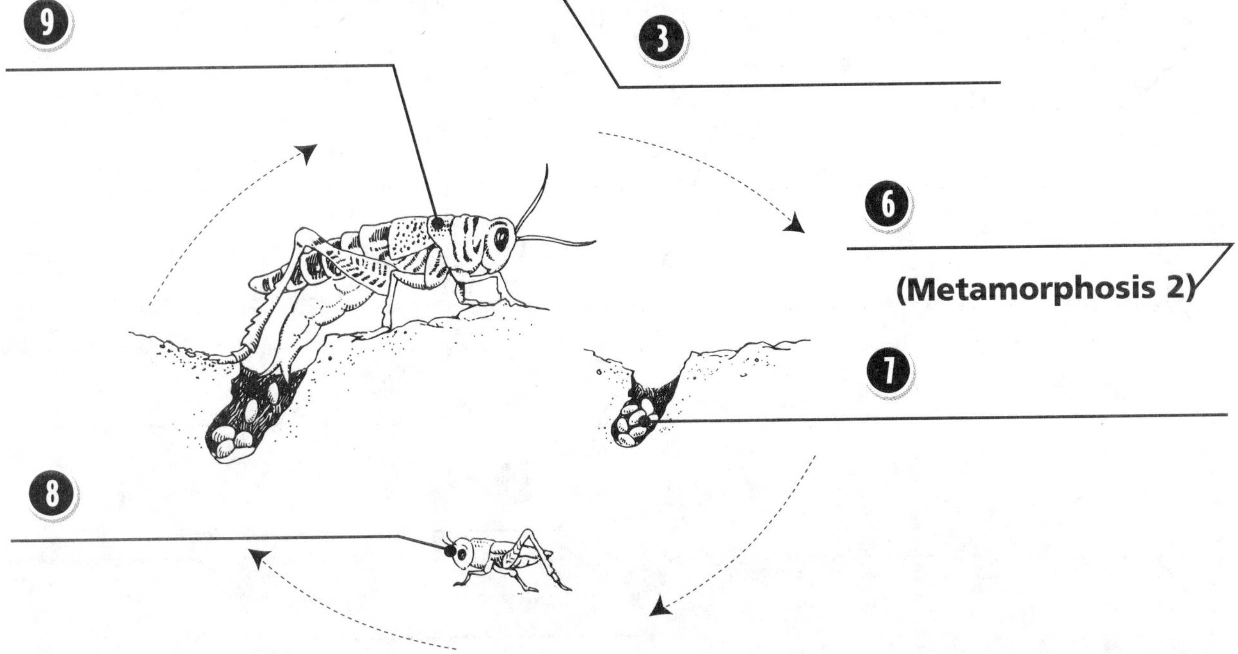

5 _____

1 _____
(Metamorphosis 1)

4 _____

2 _____

9 _____

3 _____

6 _____
(Metamorphosis 2)

7 _____

8 _____

An Echinoderm–The Sea Star

Echinoderm means "spiny skin." The animals in this phylum often have a spiny, rough exterior, radial symmetry, and tube feet. Sea stars, sea urchins, sand dollars, and sea cucumbers are members of this phylum. Use the terms in the word box to label the diagram.

ring canal	sieve plate	tube feet
radial canal	anus	digestive glands
stomach		

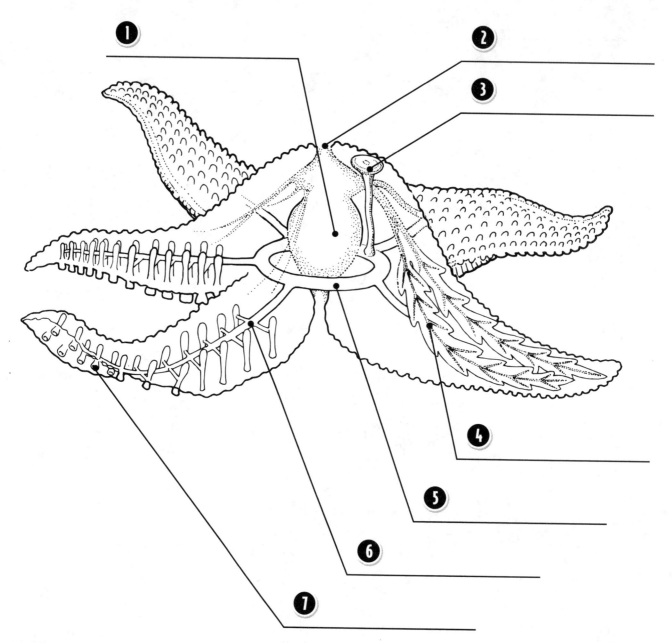

Name _____   Date _____

The Vertebrate Animals

The phylum Chordata includes those animals that are vertebrates. These bilateral animals have a
backbone made of either cartilage or bone. Their brains are protected inside a chamber of skull
bones. There are seven main classes of living vertebrates. Use the terms in the word box to label them.

bony fish   amphibian   bird
reptile   mammal   cartilage fish
jawless fish

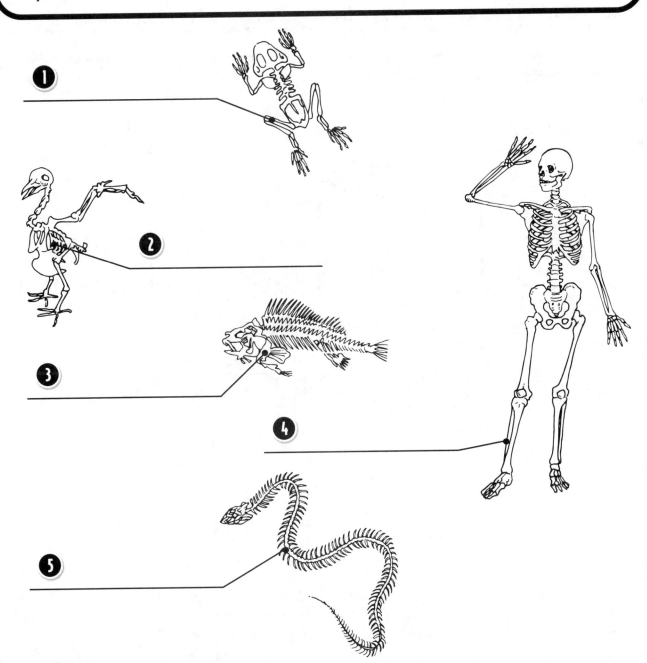

1 _____

2 _____

3 _____

4 _____

5 _____

6 These do not have bones: _____ _____

Life Science © 2004 Creative Teaching Press

Comparing the Vertebrates

Vertebrates share some characteristics and have others that are uniquely their own. Use the terms in the word box to complete the chart. Some terms are used more than once.

skeleton of cartilage	skeleton of bone	cold-blooded
warm-blooded	paired fins	jawless
gill covers	well-developed brain	dry, scaly skin
feathers	feed milk to young	leathery eggshell
gilled young; adults with lungs	hollow bones	

Cartilage Fish	Jawless Fish	Bony Fish	Amphibians
_____	_____	_____	_____
_____	_____	_____	_____
_____	_____	_____	_____

Reptiles	Birds	Mammals
_____	_____	_____
_____	_____	_____
_____	_____	_____
_____	_____	_____

Circulatory Systems of Vertebrates

The circulatory system moves blood through a body. It carries oxygen and nutrients to the cells and carbon dioxide and waste away from the cells. There are three types of circulatory systems found in vertebrates. Use the terms in the word box to complete the chart. Some terms are used more than once.

two-chambered heart	three-chambered heart	four-chambered heart
lungs	gills	ventricle
atrium	two ventricles	two atria

Fish	Reptile	Mammal
_____ _____ _____ _____	_____ _____ _____ _____	_____ _____ _____ _____

Name _____ Date _____

Characteristics of Fish

All fish can be divided into three groups: the cartilage fish, the jawless fish, and the bony fish. All are cold-blooded and breathe with gills. Other characteristics distinguish one group from the other. Use the terms in the word box to complete the chart.

skate	shark	stingray
trout	flexible skeletons	moray eel
sucker-shaped mouths	sea horse	hagfish
lamprey	no paired fins	paired fins
lobed- or ray-finned	jawless	5–7 gill slits per side
gill covers	scaled bodies	

Cartilage Fish	Jawless Fish	Bony Fish
Characteristics: _____ _____ _____	Characteristics: _____ _____ _____	Characteristics: _____ _____ _____
Examples: _____ _____ _____	Examples: _____ _____ _____	Examples: _____ _____ _____

Name _____ Date _____

Cartilage Fish

Sharks are the most well-known member of the cartilage fish. They have streamlined bodies and five different kinds of fins that help them swim. Use the terms in the word box to label the diagram.

intestine	pelvic fin	dorsal fins
kidney	eye	stomach
vertebral column	liver	gill slit
heart	caudal fin	pectoral fin

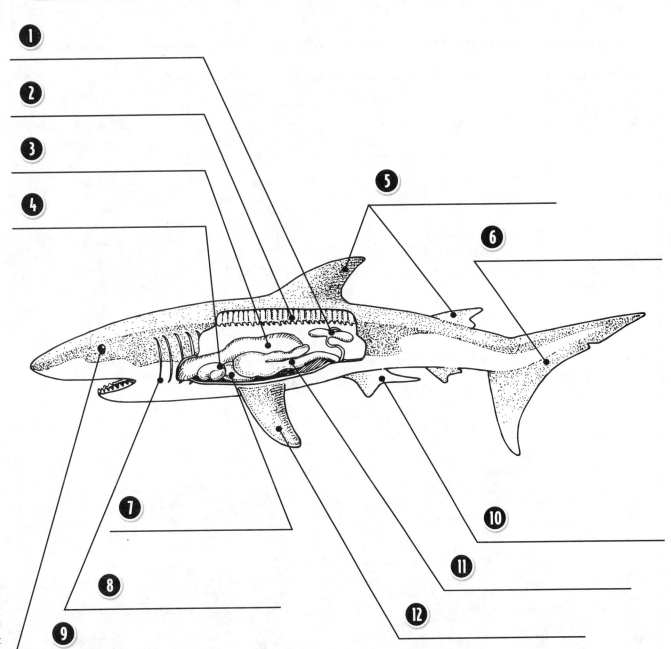

Name _____ Date _____

Bony Fish

Bony fish have skeletons made of rigid bone. Their body shapes are very diverse and well-suited to the environment in which they live. Bony fish can be divided into those with lobed fins and those with ray fins. Use the terms in the word box to label the diagrams of two ray-finned fish: a soldierfish and a perch. Some terms are used more than once.

pectoral fin	pelvic fin	anal fin
caudal fin	dorsal fin	fin supports
brain	swim bladder	anus
heart	olfactory bulb	stomach
liver	intestine	

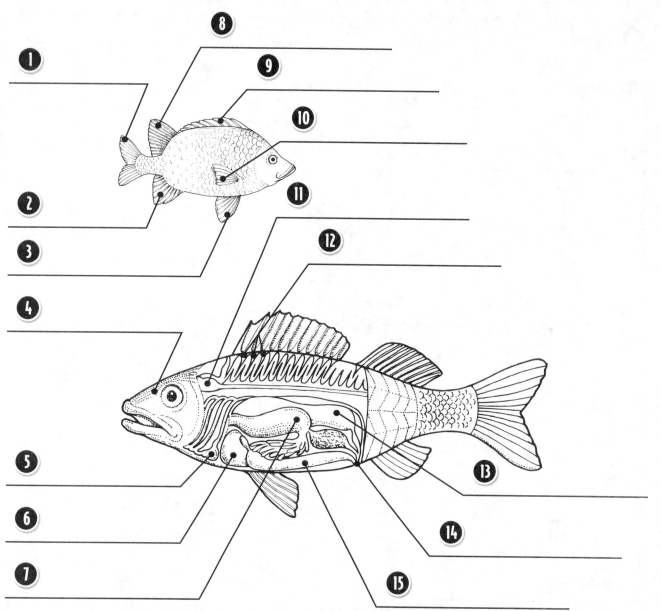

Characteristics of Amphibians and Reptiles

Amphibians and reptiles are both cold-blooded vertebrates. However, while reptiles have lungs and live largely on land, the amphibians begin their lives in the water using gills and develop lungs as they grow. Use the phrases in the word box to complete the chart.

moist skin through which water can pass
tough, dry skin with horny scales
adults breathe with lungs; young with gills
most adults are carnivorous
most hatch from eggs laid on land

have three-chambered heart
have a cloaca
young and adults breathe with lungs
rarely have scales
eggs have a leathery shell

Amphibians	Reptiles

_____

_____

_____

_____

Name _____ Date _____

An Amphibian—The Frog

Use the terms in the word box to label the diagram of a frog and a frog's mouth. Some terms are used more than once.

teeth	eye	tongue
nostril opening	mouth	foreleg
hind leg	eardrum	

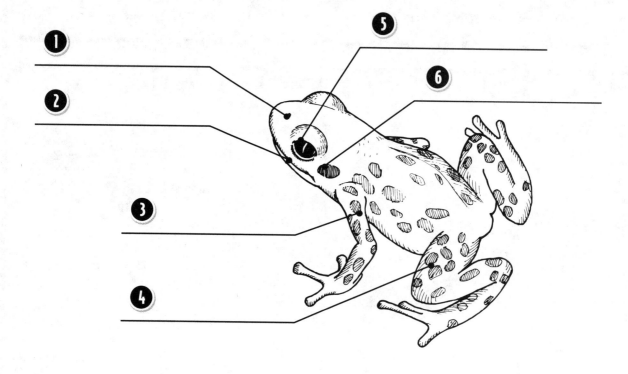

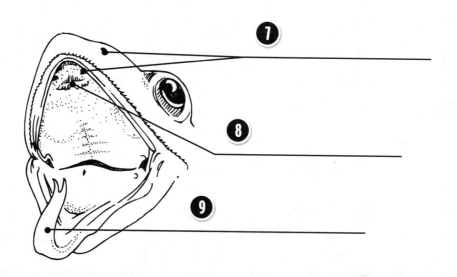

Name _____ Date _____

The Internal Organs of a Frog

Use the terms in the word box to label the diagram of a frog's internal organs.

heart	kidney	lung
anus	mesentery	stomach
liver	small intestine	large intestine

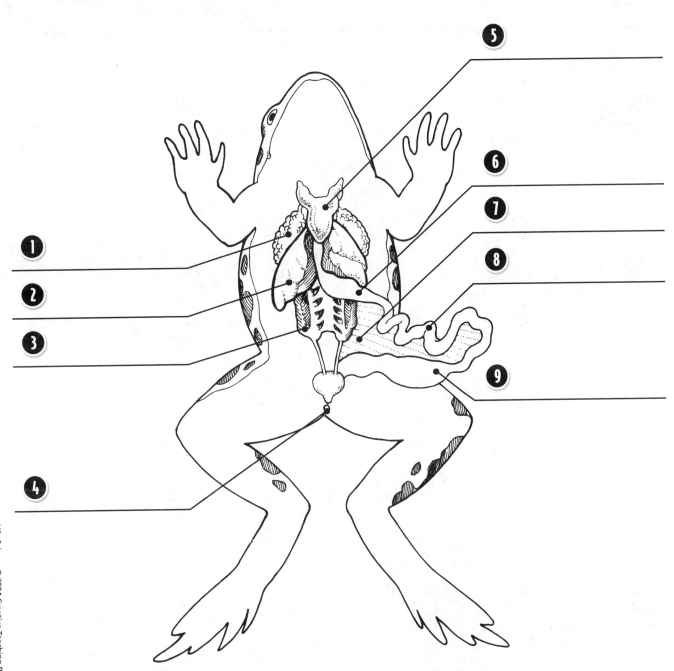

Life Cycle of an Amphibian

Most amphibians undergo metamorphosis. During this time, amphibian larvae slowly change from animals with tails and gills to animals better suited for life on land. Use the phrases in the word box to label the stages of amphibian metamorphosis.

young frog has absorbed all tail remnants
an egg mass is laid in water
tadpole frog has limbs and lungs and begins to absorb tail
embryo grows within egg

tadpole grows hind limbs
egg hatches into larvae called tadpoles

frog is mature adult

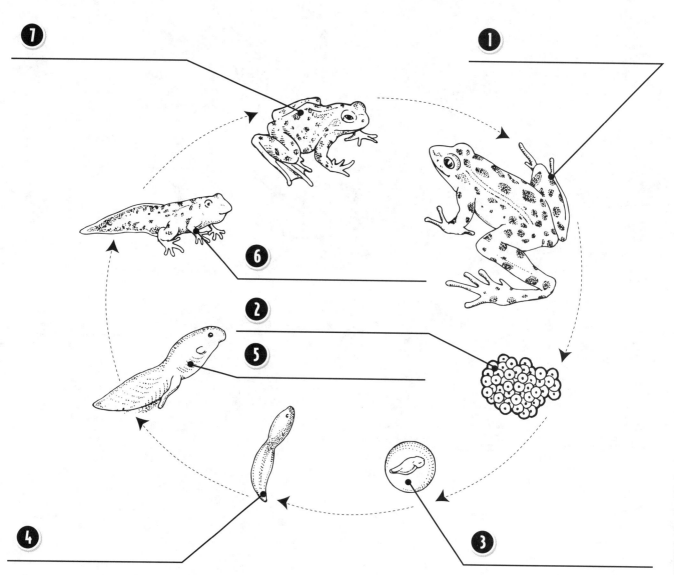

Life Science © 2004 Creative Teaching Press

A Reptile–The Alligator

Use the terms in the word box to label the diagram of a male alligator.

kidney   heart   trachea
brain   carotid artery   testes
intestine   liver   lung
stomach   cloaca   posterior vena cava
esophagus

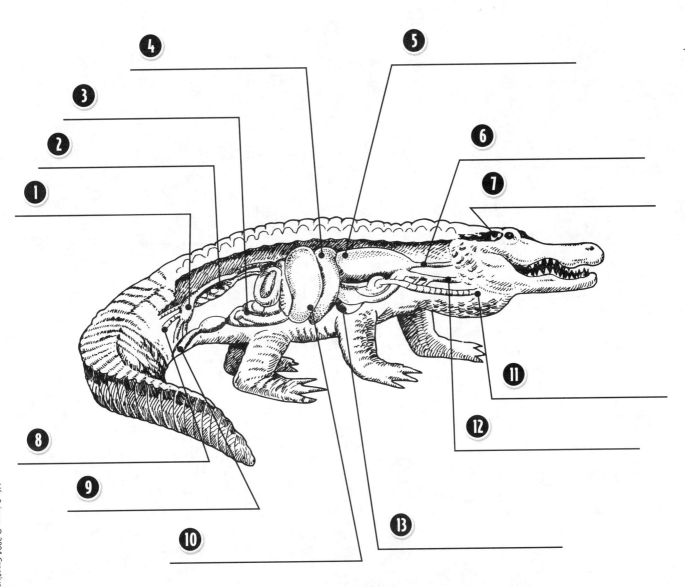

Name _____ Date _____

The Head of a Venomous Snake

Many reptiles, especially the snake, have parts used to seek out, then stun or kill prey. Use the terms in the word box to label the diagram of a venomous snake's head.

eye	fang	pit
teeth	nostril	tongue
scales	glottis	venom sac

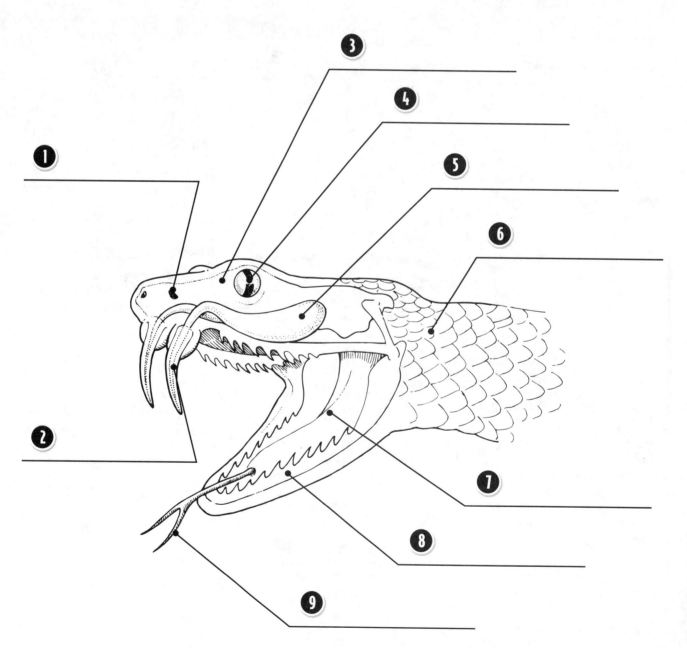

Name _____ Date _____

Characteristics of Birds and Mammals

Birds and mammals are both warm-blooded vertebrates. Both have characteristics that make them unique within the animal kingdom. Use the terms in the word box to complete the chart. Some terms are used more than once.

> have four-chambered hearts
> develop from embryos in eggs outside the mother's body
> have hollow bones
> have toothless, lightweight jaws
> most have specialized teeth
> most give birth to live young
>
> have hair
> have feathers
> raise young on milk
> have wings
> have mammary glands

Birds	Mammals
_____	_____
_____	_____
_____	_____
_____	_____
_____	_____

Parts of a Bird

Nearly all birds can fly. Even those that do not have the same specialized body parts that make birds unique. Use the terms in the word box to label the diagram.

crown	belly	tail feathers
breast	primary feathers	beak
throat	back	talons
secondary feathers		

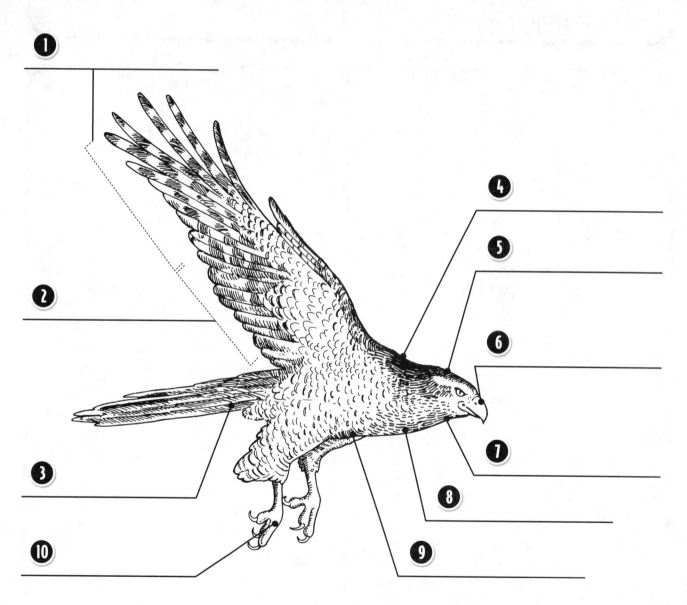

A Bird's Internal Organs

Use the terms in the word box to label the diagram of a bird's internal organs.

lung	eye	wing bone
liver	heart	spinal cord
esophagus	beak	crop
gizzard	sternum	rectum
trachea	kidney	

① _____

② _____

③ _____

④ _____

⑤ _____

⑥ _____

⑦ _____

⑧ _____

⑨ _____

⑩ _____

⑪ _____

⑫ _____

⑬ _____

⑭ _____

Name _____   Date _____

Bird Beaks

A bird's beak shows amazing adaptation. Beaks are specialized by size and shape to assist birds as they capture and manipulate food. Use the terms in the word box to label each illustration.

tears flesh of animals
cracks nuts and seeds
scoops up water and fish

sucks nectar from flowers
traps insects in midair
stabs small fish

Pelican	Cockatoo	Eagle
❶ _____	❷ _____	❸ _____
Hummingbird	Heron	Sparrow
❹ _____	❺ _____	❻ _____

Name _____ Date _____

Fowl Feet

A bird's feet are also specialized body parts that give clues about its habits or environment. Use the terms in the word box to complete the chart.

for perching on branches	for catching prey
for swimming	for grasping in order to climb
for wading in mud	goose
thrush	hawk
woodpecker	coot

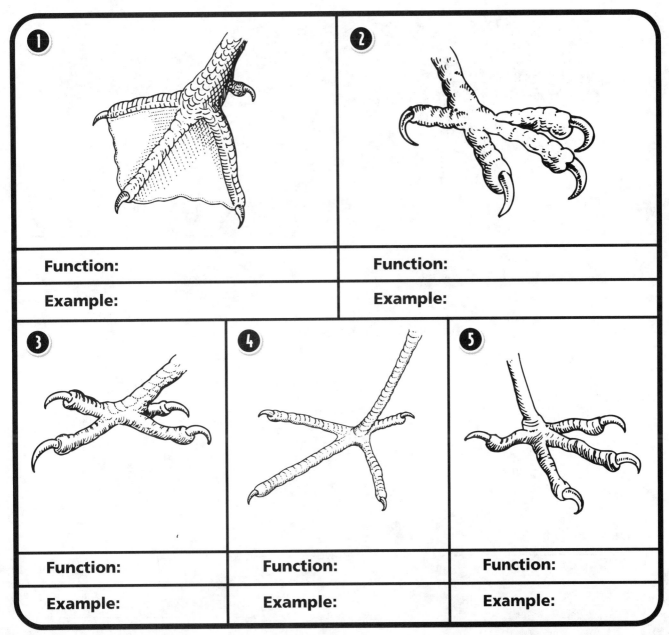

1

Function:

Example:

2

Function:

Example:

3

4

5

Function:

Example:

Function:

Example:

Function:

Example:

Name _____   Date _____

Birds of a Feather

The wings and body of a bird are covered with different types of feathers. Each type has a certain purpose. Use the terms in the word box to label each diagram. Some terms are used more than once.

> body feathers
> down feathers
>
> primary flight feathers
> secondary flight feathers
>
> coverts

1 _____

Parts of a Wing

2 _____

3 _____

Types of Feathers

 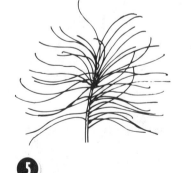 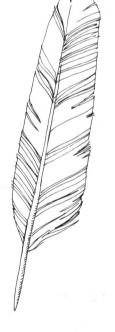

4 _____

5 _____

6 _____

Mammals

Mammals can be divided into three groups: the monotremes, the marsupials, and the placentals. Use the terms in the word box to classify each illustration.

Monotremes: These mammals lay eggs, a trait unique to this group of mammals alone.
Marsupials: These mammals give birth to live young in a very undeveloped state. Once born, they continue to develop in a pouch in the mother's abdomen.
Placentals: These young stay in their mother's bodies for a relatively long time, nourished by placenta tissue as they develop.

1 _____   2 _____   3 _____

4 _____   5 _____   6 _____

7 _____   8 _____   9 _____

Name _____ Date _____

Mammal Locomotion

Mammals move around in a number of ways. Most walk or run, many swim, and some even fly. While many mammals can move about in more than one way, most are designed to move about most easily in one manner. Use the terms in the word box to label each animal with its primary way of moving.

fly	walk and run	swim

1 _____

2 _____

3 _____

4 _____

5 _____

6 _____

7 _____

8 _____

9 _____

Life Science © 2004 Creative Teaching Press

Name _____ Date _____

The Science of Ecology

Ecology is the study of how plants and animals interact with each other and their physical environment. Match each term in the word box to its definition.

```
biosphere                physical environment      population
ecosystem                niche                     food web
biological environment   energy                    habitat
diversity
```

1 The _____ includes solar light and heat, moisture, wind, oxygen, carbon dioxide, and nutrients.

2 The _____ includes the plants and animals in a habitat.

3 The _____ refers to the thin layer of the earth in which living organisms are found.

4 An _____ is a single community of plants, animals, and their physical environment.

5 _____ is transferred from the sun, to the producers, to the consumers, and to the decomposers back into the physical environment.

6 A _____ is how energy is transferred through an ecosystem from organism to organism.

7 A _____ is a group of organisms of the same kind living in the same place at the same time.

8 A _____ is the place where particular plants or animals live.

9 A _____ is the particular role that an organism serves in a habitat.

10 _____ refers to how many different species are found within a particular habitat or ecosystem.

Name _____ Date _____

Producers, Consumers, and Decomposers

Organisms are either producers or consumers, depending on the source of their energy. A third group, decomposers, get their energy by breaking down the remains of other organisms. Use the terms in the word box to classify each plant or animal.

| producer | consumer | decomposer |

1 _____

2 _____

3 _____

4 _____

5 _____

6 _____

7 _____

8 _____

9 _____

Name _____ Date _____

Carnivores, Herbivores, and Omnivores

An herbivore is an organism that eats plants. A carnivore is an organism that eats animals. An omnivore is an organism that eats both plants and animals. Use the terms in the word box to label each illustration.

| herbivore | carnivore | omnivore |

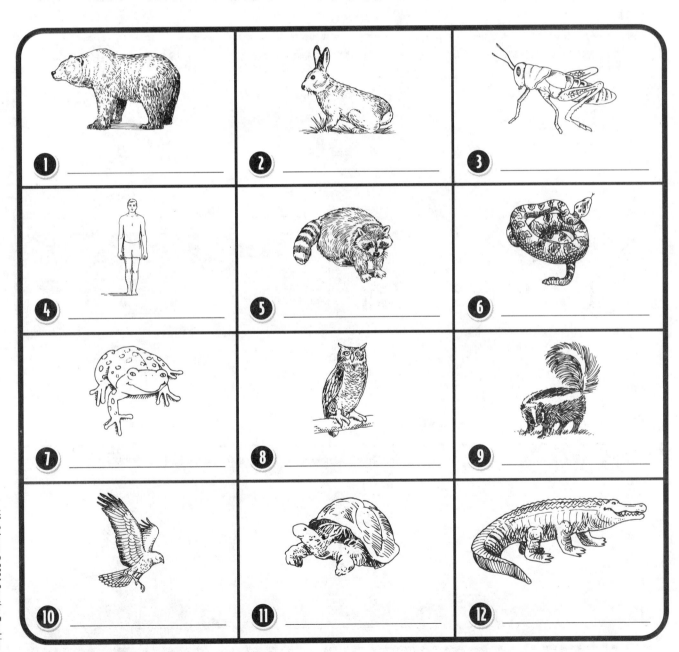

1 _____

2 _____

3 _____

4 _____

5 _____

6 _____

7 _____

8 _____

9 _____

10 _____

11 _____

12 _____

Parasites

Parasitism describes when two organisms live together, one drawing its nourishment at the expense of the other. Parasites normally do not kill their hosts the way predators do. However, the host is weakened by the presence of the parasite. Use the terms in the word box to match the parasite to its host.

attaches to tree tissue	lives in intestine	enters the bloodstream
lives off of other insects	burrows into skin	blood-sucking insect
blood-sucking segmented worms	feeds off of forest plants	attaches to fish

Parasite: lamprey

1 _____

Parasite: tapeworm

2 _____

Parasite: blood fluke

3 _____

Parasite: mistletoe

4 _____

Parasite: tick

5 _____

Parasite: nematodes

6 _____

Parasite: mosquito

7 _____

Parasite: leech

8 _____

Parasite: Indian Pipe

9 _____

Plant Succession

A community grows through slow and gradual change over time. As one species replaces or dominates others, the community changes. This is called succession. Use the phrases in the word box to label each illustration. Number the illustrations to show the order of succession.

> water collects and forms a freshwater pond
> bog fills in to form a meadow
>
> sediments in water create a bog
> trees take root to form a forest

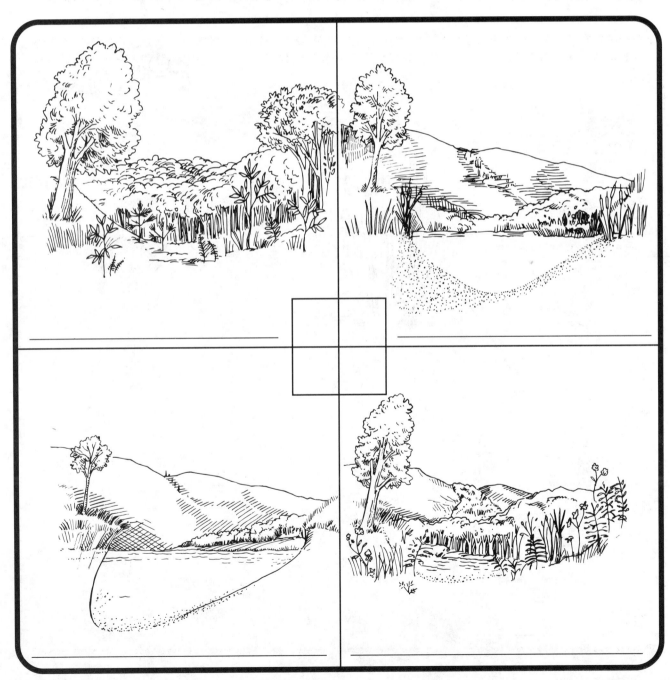

Food Chains

The transfer of energy through organisms forms a food chain. Use the terms in the word box to label the diagram. Then draw arrows to show the transfer of energy through the food chain. One arrow is drawn for you.

human	trout	frog
grasshopper	grass	sun

_____

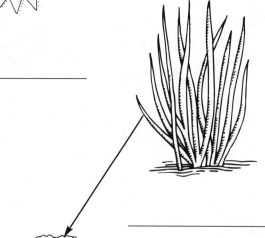

_____

_____

_____

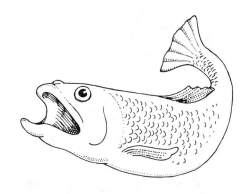

_____

Name _____ Date _____

Food Webs

Consumers do not eat just one organism. Often, organisms feed off of a variety of producers or other consumers and share these food sources with other consumers. Interconnected food chains form food webs. Use the terms in the word box to label the diagram. Then use arrows to show the many ways energy transfers from one organism to another.

grasshopper	deer	frog
trout	grass	wolf
mouse	snake	raccoon

_____

_____

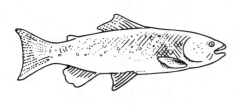

_____

_____

_____

_____

_____

Adaptations

Over time, organisms have adapted to their environment in order to survive. Match each term in the word box to the animal it describes.

> stems of plants store water in dry conditions
> hibernate to survive low food supplies in winter
> have blubber to insulate body against cold
> have special hemoglobin for high altitude living
> have densely packed fur so skin does not get wet
> have specialized digestive tract to digest plant cellulose

①

_____

_____

②

_____

_____

③

_____

_____

④

_____

_____

⑤

_____

_____

⑥

_____

_____

Life Science © 2004 Creative Teaching Press

Camouflage

Camouflage is the concealment of animals with their surroundings. Match each animal to the description of its camouflage technique.

snowshoe hare	owl butterfly	walking stick
moth	caterpillar	walking leaf
halibut	deer	

1 _____ The colors and patterns on the wings of this animal match the bark where it rests during the day.

2 _____ This animal resembles the twigs where it feeds, hiding it from other predators.

3 _____ This insect displays mimicry. The spots on its wings look like the eyes of a predatory bird, scaring off other predators.

4 _____ This larval stage of the swallowtail butterfly also has large spots resembling eyes, making it look frightening to birds and other predators.

5 _____ This mammal changes color during the summer and winter so that it is able to blend in with the snow or the vegetation of its habitat.

6 _____ This insect has enlargements on its legs and abdomen that make it resemble the plants it feeds on.

7 _____ This fish has a flattened body and coloration that allow it to blend in with the sea floor.

8 _____ The young of this mammal have spotted fur to allow them to blend in with the dappled sunlight of the forest floor.

Defense Mechanisms

Each of these animals has a special defense adaptation. Use the terms in the word box to identify the animal and its defense mechanism.

skunk	pretends it is dead
sprays offensive odor	opossum
releases quills upon contact	retreats into a shell
porcupine	runs and kicks with powerful legs
ostrich	tortoise
octopus	releases a cloud of ink

Animal:	**Animal:**	**Animal:**
Defense:	**Defense:**	**Defense:**
Animal:	**Animal:**	**Animal:**
Defense:	**Defense:**	**Defense:**

Name _____ Date _____

The Deserts

The desert biomes are regions that get less than ten inches of rainfall and have a high average temperature and where evaporation exceeds precipitation. Use the numbered terms in the word box to label the diagram.

1. desert floor radiates heat back into atmosphere at night
2. lack of vegetation causes erosion during rainstorms
3. winds erode rock into interesting formations
4. long root systems collect as much rainfall as possible
5. thick stems store water
6. tiny leaves or hairy surfaces minimize evaporation
7. animals obtain water from food they eat
8. reptiles can control the amount of heat they produce or release

Name _____ Date _____

The Tundra

The tundra is characterized by harsh winters, low average temperatures, little snow or rainfall, and a short summer season. Use the numbered terms in the word box to label the diagram.

1. has a layer of permafrost
2. poor drainage forms ponds and bogs
3. many lichen grow in this environment
4. plants are low growing with greater root mass
5. birds migrate to warmer climates in winter
6. animals have fur that reduces heat loss
7. animals have blubber layer to reduce heat loss

Name _____ Date _____

Savannas and Grasslands

The savannas and grasslands are biomes with grasses and few trees. They have many similarities, but they differ in their location, climate, and wildlife. Use the phrases in the word box to complete the chart. Some phrases are used more than once.

> contains grasses, shrubs, and few small trees
> rainfall averages 35–60 inches (90–150 cm) a year
> rainfall averages 10–40 inches (25–100 cm) a year
> vegetation maintained by periodic fire
> large grazing animals such as elephants, wildebeest, and zebra
> found between deserts and tropical forests
> found between deserts and temperate forests
> large grazing animals such as bison and antelope

American Grassland	African Savanna
_____	_____
_____	_____
_____	_____
_____	_____

Name _____ Date _____

Tropical Rain Forests

Tropical rain forests are hot and humid throughout the year. They get more than 80 inches (2.032 meters) of rain a year and temperatures usually remain around 70°–85°F (21–29°C). Match each term to its definition.

vegetation
nutrients
canopy
microorganisms

more than half
understory
surface

2–5 percent
emergent
forest floor

1 _____ Scientists estimate that this is the amount of the world's species that are found in tropical rain forests.

2 _____ The heavy rainfall and the old age of the rain forest soils means that it lacks this.

3 _____ The roots of most rain forest trees can be found in this area of the soil.

4 _____ These live in the soil and quickly break down dead animal and plant material into nutrients that can be used by living organisms.

5 _____ Because of how dense this is, the rain forest can be divided into layers. Each layer contains plants and animals that have adapted to the amount of light, temperature, and availability of nutrients.

6 _____ This layer towers above all others and consists of giant trees.

7 _____ This layer is formed by flat-crowned trees and acts like a giant sun and rain umbrella.

8 _____ The canopy catches most of the sun's rays, only allowing this amount to slip through to the forest floor.

9 _____ This shadowy layer is made up of small trees and miniature woody plants. They have large leaves to catch all available light.

10 _____ In this bottom layer, life is much different than at the top. The air is still and the temperature and humidity remain constant. Vegetation is sparse.

Name _____ Date _____

Coniferous Forests

The coniferous forests are found in regions in which a cold, dry season alternates with a cool, rainy season. Match each term in the word box to its definition.

peat moss   montane   boreal
taiga   temperate rain forest   conifers
logging   pine barrens

1 _____ These are the primary producers in a coniferous forest.

2 _____ These forests are found in Northern Europe, Asia, and North America where cold lakes and streams are a part of the landscape.

3 _____ Another name for a boreal forest, this term means "swamp forest."

4 _____ This plant dominates in boreal forests where drainage is poor and the standing water creates bogs.

5 _____ These forests are found in areas of coastal New Jersey with sandy, nutrient-poor soil.

6 _____ This forest stretches along the coast from Alaska to Northern California and is home to some of the world's tallest trees.

7 _____ These coniferous forests are found in the great mountain ranges of the Northern Hemisphere. Fir, spruce, and pine dominate in these areas.

8 _____ Throughout the centuries, this industry has destroyed large portions of coniferous forests.

Name _____ Date _____

Wetland Biomes

Wetlands are found where the land is periodically or permanently flooded. These areas have special soils that have developed over time under these waterlogged conditions. The plants and animals that are found in these biomes are adapted to these conditions. Match each term in the word box to its description.

marsh	swamp	bog
salt marsh	mangrove swamp	endangered
impermanent	emergents	floating aquatics
submergents	migration	

1 _____ This is a coastal wetland that is affected by tides and periodically covered with ocean water. It is dominated by soft-stemmed plants such as grasses, rushes, and sedges.

2 _____ This is a wetland found inland and contains freshwater. It is dominated by soft-stemmed plants such as grasses, rushes, and sedges.

3 _____ This describes the state of wetland biomes. Changes in precipitation, accumulating sediments, and rising sea levels all affect and change wetland biomes.

4 _____ This describes many of the species that are found in wetland biomes. The changing nature of wetlands and the impact of human activities lead to this problem.

5 _____ These are plants that live partly in and partly out of water, such as cattails and grasses.

6 _____ These are plants that float on the water's surface, such as water lilies and water hyacinth.

7 _____ This is a freshwater wetland dominated by woody plants, such as shrubs and trees.

8 _____ This is a wetland that grows on thick mats of peat, a substance made of partially decayed plants and animals.

9 _____ These are plants that grow entirely underwater. Elodea and eel grass are examples.

10 _____ This is a coastal wetland that is affected by the tides and filled with salty ocean water. Mangrove trees are the dominant plant material in this wetland.

11 _____ Many waterfowl and shorebirds nest in or stop over to feed in wetlands during this seasonal process.

Name _____ Date _____

Tidal Zone Biomes

Along the coastline of oceans and continents, there are diverse animal and plant species. Each species has special adaptations that allow them to thrive in a particular tidal zone. Use the terms in the word box to label the diagram.

upper intertidal splash zone	shrimp	limpets
lower intertidal zone	mussel	subtidal zone
mid-intertidal zone	sea anemone	crab
sea star	octopus	barnacles

Zone: _____

Species: _____

_____

Zone: _____

Species: _____

_____

Zone: _____

Species: _____

_____

Zone: _____

Species: _____

_____

Coral Reefs

Coral reefs rise up from the ocean floor. Their structure is created by corals. Corals are animals that produce limestone as they grow. Successive generations of corals live upon the limestone remains of their ancestors. Use the terms in the word box to label the diagrams.

lagoon	reef flat	fringing reef
barrier reef	reef crest	atoll
fore-reef		

Types of Coral Reefs

_____	_____	_____
Forms an edge around a shore or continent.	Forms close to the shore, but separated by a lagoon.	Occurs as a large ring shape in open ocean.

Zones of a Coral Reef

Life Science © 2004 Creative Teaching Press

Answer Key

Life Processes (page 5)

1. reproducing
2. growing
3. getting energy
4. responding to change
5. getting rid of waste
6. using energy
7. getting energy
8. using energy
9. reproducing

The Carbon Dioxide-Oxygen Cycle (page 6)

1. carbon dioxide
2. marine algae
3. producers
4. photosynthesis
5. consumers
6. fossil fuels
7. decomposers
8. aerobic
9. geologic activity
10. oxygen

The Nitrogen Cycle (page 7)

1. amino acids
2. atmosphere
3. lightning
4. bacteria
5. animals
6. decomposition
7. leaching
8. legumes
9. nitrification
10. ammonia

The Five Kingdoms (page 8)

1. Monera   Common Name: bacteria
2. Protista   Common Name: single-celled organisms
3. Fungi   Common Name: molds, mushrooms, lichen
4. Plantae   Common Name: plants
5. Animalia   Common Name: animals

Classification of Living Things (page 9)

Kingdom Monera
bacillis
salmonella
streptococcus
spirochetes
lactobacillis

Kingdom Protista
amoeba
paramecium
algae
euglena
protozoa

Kingdom Fungi
lichens
bread molds
mushrooms
truffles
yeast

Kingdom Plantae
conifers
monocots
ferns
horsetails
dicots

Kingdom Animalia
insects
mammals
amphibians
birds
reptiles

Plant or Animal? (page 10)

Plant
living organisms
formed from cells
cells have chlorophyll
makes own food
has limited movement
reproduces its own kind
depends on sun's energy

Animal
living organisms
formed from cells
cells have no chlorophyll
moves from place to place
reproduces its own kind
depends on sun's energy
obtains food from outside sources

The Structure of Bacteria (page 11)

1. pilus
2. capsule
3. DNA
4. ribosomes
5. cytoplasm
6. plasma membrane
7. cell wall
8. bacterium flagellum

Bacterial Shapes (page 12)

1. bacilli; rod-shaped
2. spirochetes; corkscrew-shaped
3. cocci; spherical
4. bacilli
5. bacilli
6. cocci
7. spirochete

Classifications of Bacteria (page 13)

1. aerobic
2. autotrophs
3. anaerobic
4. beneficial
5. cyanobacteria
6. chemoautotrophs
7. heterotrophs
8. pathogenic

What It Takes to Be a Protistan (page 14)

1. long flagellum
2. contractile vacuole
3. chloroplast
4. pellicle
5. mitochondrion
6. Golgi body
7. endoplasmic reticulum
8. nucleus
9. light-sensitive spot

The Paramecium (page 15)

1. micronucleus
2. macronucleus
3. gullet
4. cilia
5. food vacuole
6. contractile vacuole
7. trichocysts
8. contractile vacuole

Moving Like an Amoeba (page 16)

1. nucleus
2. food vacuole
3. false foot
4. water vacuole
5. nucleus
6. cell membrane
7. cytoplasm

Red, Brown, and Green Algae (page 17)

1. bladder
2. blades
3. stipe
4. holdfast
5. holdfast
6. stipe
7. bladder
8. blades

A Fungus Among Us (page 18)

1. mushroom
2. puffballs
3. bracket fungi
4. boletus
5. crusty lichen
6. shrubby lichen
7. mold
8. yeast

The Nature of Fungi (page 19)

1. penicillin
2. mold
3. fungus
4. lichen
5. budding
6. yeast
7. fungi
8. toadstool
9. chlorophyll
10. parasite

Parts of a Mushroom (page 20)

1. cap
2. gills
3. membrane
4. ring
5. stalk
6. mycelium

Growing Up Mushroom (page 21)

1. full-grown
2. half-grown
3. button stage
4. pinhead stage
5. gills
6. ring
7. veil broken to form ring
8. veil
9. gills inside veil
10. spawn

Dividing to Multiply (page 22)

1. nucleus moves toward dividing end of yeast cell
2. bud forms as nucleus begins to divide
3. cell wall begins to form new cell
4. cells fully split with nucleus in each cell
5. cytoplasm
6. bud
7. cell wall
8. nuclei (plural for nucleus)
9. vacuole

The World of Plants (page 23)

1. moss
2. conifer
3. monocot
4. dicot
5. fern
6. deciduous tree

A Typical Plant Cell (page 24)

1. nucleus
2. mitochondrion
3. cell wall
4. cell membrane
5. chloroplast
6. ribosomes
7. endoplasmic reticulum
8. cytoplasm
9. vacuole

Functions within a Plant Cell (page 25)

1. cell wall
2. vacuole
3. chloroplast
4. endoplasmic reticulum
5. cytoplasm
6. organelle
7. ribosome
8. mitochondrion
9. nucleus
10. cell membrane

Photosynthesis (page 26)

1. sunlight
2. glucose
3. Chlorophyll
4. carbon dioxide
5. chloroplasts
6. Water
7. oxygen
8. carbohydrates
9. cellulose
10. leaves
11. photosynthesis
12. autotrophs
13. heterotrophs

Mosses (page 27)

1. leaves
2. seta
3. stems
4. calyptra
5. foot
6. capsules
7. sporophyte
8. gametophyte
9. rhizoid

Ferns (page 28)

1. frond
2. leaflets
3. sorus
4. spores
5. fiddlehead
6. roots
7. rhizome

Horsetails (page 29)

1. spore-bearing structures
2. scalelike leaves
3. hollow stems
4. ribs
5. rhizomes

Gymnosperms (page 30)

1. seed
2. spores
3. cone
4. needles
5. roots
6. trunk

Conifer Needles, Scales, and Cones (page 31)

1. shell around seed kernel
2. young cone
3. flat needles
4. seed kernel
5. cone scale
6. mature cone
7. needles in pairs
8. needles in threes
9. needles in fives
10. scalelike needles

Life Cycle of a Conifer (page 32)

1. cone
2. gametophyte
3. pollen grains
4. seedling
5. embryo
6. male cones
7. wind currents
8. fertilization
9. pollination
10. female cones

The Angiosperms (page 33)

1. flower
2. leaf
3. seeds
4. fruit
5. stem
6. root

Monocots and Dicots (page 34)

Monocot
one seed leaf
parallel leaf veins
flower parts in threes
branching fibrous roots
scattered bundles

Dicot
two seed leaves
netlike leaf veins
flower parts in fours or fives
central taproot
bundles in a ring

Flowering Plant Parts (page 35)

1. pollinator
2. ovule
3. fruit
4. petal
5. bud
6. pollen
7. leaf
8. seed
9. root
10. sepal
11. stem
12. calyx

Root Systems (page 36)

1. stems and leaves
2. prop roots
3. crown
4. lateral root
5. taproot system
6. root tip
7. fibrous roots
8. root hair cell

Inside a Root (page 37)

1. vascular cylinder
2. primary phloem
3. primary xylem
4. root hair
5. root tip
6. epidermis
7. root cap
8. lateral root
9. root cortex

Underground Stems (page 38)

1. leaf
2. bud
3. bud
4. root
5. stem
6. root
7. rhizome
8. tuber
9. bulb

Stems Above Ground (page 39)

1. blade
2. petiole
3. bud
4. node
5. node
6. blade
7. vascular bundles
8. sheath
9. pith
10. vascular bundles
11. cortex
12. ground tissue

Woody or Herbaceous Stems (page 40)

1. phloem
2. cork
3. pith
4. xylem
5. cortex
6. cambium
7. bark
8. cambium
9. phloem
10. cortex
11. xylem
12. pith
13. epidermis

Really Big Stems (page 41)

1. leaves
2. branches
3. trunk
4. roots
5. bark
6. phloem
7. cambium
8. xylem
9. sapwood
10. heartwood

A Look on the Inside (page 42)

1. bark
2. vascular cambium
3. xylem
4. phloem
5. heartwood
6. sapwood
7. bark

As a Tree Grows (page 43)

1. bud-scale scar
2. branch
3. flower bud
4. terminal bud
5. node
6. internode
7. leaf bud
8. leaf scar

Looking at Leaves (page 44)

1. lobe
2. blade
3. petiole
4. leaflet
5. compound leaf
6. leaf margin
7. simple leaf
8. veins

Characteristics of Leaves, Part One (page 45)

1. simple
2. palmate
3. pinnate
4. v-shaped
5. rounded
6. flat
7. heart
8. uneven
9. compound

Characteristics of Leaves, Part Two (page 46)

1. saw-toothed
2. smooth
3. double saw-toothed
4. lobed
5. parallel
6. palmate
7. pinnate
8. opposite
9. alternate

A Closer Look (page 47)

1. stomata
2. guard cells
3. waxy layer
4. vein
5. epidermis
6. vein
7. stomata
8. palisade layer
9. spongy layer
10. guard cell

Seed-Producing Parts of a Flower (page 48)

1. anther
2. filament
3. stamen
4. stigma
5. style
6. ovary
7. carpel
8. ovule
9. receptacle

The Process of Pollination (page 49)

1. corolla
2. stamens
3. anther
4. filament
5. pollen grains
6. carpel
7. ovary
8. stigma
9. style
10. perfect

Monocots (page 50)

1. embryo
2. endosperm
3. cotyledon
4. epicotyl
5. hypocotyl
6. radicle
7. seed and fruit coats
8. embryo

Dicots (page 51)

1. hilum
2. radicle
3. epicotyl
4. hypcotyl
5. embryo
6. cotyledons
7. seed coat

Growth of a Monocot Plant (page 52)

1. primary leaf
2. seed coat
3. primary root
4. leaf sheath
5. branch root
6. first internode
7. first foliage leaf
8. prop root
9. fibrous roots

Growth of a Dicot Plant (page 53)

1. hypocotyl
2. cotyledons
3. seed coat
4. primary root
5. primary leaf
6. withered cotyledons
7. foliage leaf
8. node
9. point where cotyledons were attached
10. branch roots

Members of the Animal Kingdom (page 54)

1. Cnidaria
2. Platyhelminthes
3. Echinodermata
4. Arthropoda
5. Annelida
6. Chordata
7. Mollusca
8. Porifera

A Typical Animal Cell (page 55)

1. cytoplasm
2. nucleolus
3. mitochondrion
4. cell membrane
5. endoplasmic reticulum
6. nucleus
7. vacuole
8. ribosomes
9. lysosome

Functions within an Animal Cell (page 56)

1. nucleolus
2. endoplasmic reticulum
3. nucleus
4. cytoplasm
5. mitochondria
6. lysosome
7. vacuole
8. cell membrane

Body Symmetry (page 57)

1. bilateral
2. bilateral
3. radial
4. asymmetrical
5. asymmetrical
6. radial
7. bilateral
8. asymmetrical
9. bilateral
10. radial
11. bilateral
12. bilateral

The Sponges (page 58)

1. osculum
2. epithelial cell
3. collar cell
4. spicules
5. ostium
6. flagellum
7. nucleus
8. water flow

The Hydra (page 59)

1. mouth
2. tentacle
3. gastrovascular cavity
4. bud
5. foot
6. nematocyst
7. ectoderm
8. mesoglea
9. endoderm
10. ovary

Flatworms (page 60)

1. brain
2. ovary
3. eyespot
4. branching gut
5. pharynx
6. nerve cord
7. testis
8. protonephridia

Mollusks—A Snail (page 61)

1. gill
2. anus
3. mantle cavity
4. radula
5. mouth
6. excretory organ
7. heart
8. digestive gland
9. stomach
10. mantle
11. shell
12. foot

Mollusks—A Clam (page 62)

1. muscle
2. mantle
3. gill
4. mouth
5. foot
6. muscle
7. inhalant siphon
8. exhalant siphon
9. shell
10. palps

Mollusks—A Squid (page 63)

1. digestive gland
2. brain
3. radula
4. arm
5. jaw
6. tentacle
7. siphon
8. anus
9. ink sac
10. esophagus
11. kidney
12. stomach
13. internal shell
14. reproductive organ
15. mantle
16. accessory heart
17. heart
18. gill

Segmented Worms (page 64)

1. hearts
2. mouth
3. clitellum
4. ventral blood vessel
5. dorsal blood vessel
6. setae
7. esophagus
8. crop
9. gizzard
10. intestine
11. anus

The Arthropods (page 65)

Diplopoda
rounded
segmented body
two pairs of legs per segment

Chilopoda
flat
segmented body
one pair of legs per segment

The Crustaceans
hard
flexible exoskeleton
gills
two pairs of antennae
two body sections

The Arachnids
two body sections
no antennae
four pairs of legs

The Insects
three body sections
one pair of antennae
three pairs of legs

A Crustacean—The Crayfish (page 66)

1. cephalothorax
2. abdomen
3. eye
4. maxillipeds
5. tailfan
6. swimmerets
7. walking legs
8. antennae
9. cheliped

A Close Look at Arachnids (page 67)

1. abdomen
2. heart
3. cephalothorax
4. eye
5. poison gland
6. pedipalp
7. chelicera
8. mouth
9. book lung
10. silk gland
11. ovary
12. spinnerets
13. anus

In the Right Order (page 68)

Coleoptera
stag beetle
ladybug

Diptera
mosquito
housefly

Hemiptera
giant water bug
back swimmer

Lepidoptera
butterfly
silkworm moth

Hymenoptera
honeybee
ant

Odonata
dragonfly
damselfly

Orthoptera
grasshopper
praying mantis

The Honeybee (page 69)

1. abdomen
2. thorax
3. head
4. heart
5. rectum
6. stinger
7. stinger apparatus
8. simple eyes
9. antennae
10. compound eyes
11. mandible
12. ganglia
13. honey stomach
14. pollen basket
15. intestine

Life Cycle of a Bee (page 70)

1. egg laid by queen
2. grub fed by worker
3. full-grown bee grub
4. grub sealed in its cell
5. grub becomes a pupa
6. young adult leaves cell

A Grasshopper (page 71)

1. antennae
2. compound eye
3. front wing
4. ovipositor
5. knee joint
6. walking legs
7. spiracle
8. hind wing
9. jumping leg
10. femur
11. head
12. thorax
13. abdomen

Butterflies and Moths (page 72)

1. forewing
2. antennae
3. compound eye
4. proboscis
5. legs
6. abdomen
7. hind wing

Resting
wings held upright or flat
wings folded

Antennae
smooth with end knobs
feathered with no knobs

Development
chrysalis
cocoon

Metamorphosis (page 73)

1. Complete
2. egg
3. larva
4. pupa
5. adult
6. Incomplete
7. egg
8. nymph
9. adult

An Echinoderm—The Sea Star (page 74)

1. stomach
2. anus
3. sieve plate
4. digestive glands
5. ring canal
6. radial canal
7. tube feet

The Vertebrate Animals (page 75)

1. amphibian
2. bird
3. bony fish
4. mammal
5. reptile
6. jawless fish, cartilage fish

Comparing the Vertebrates (page 76)

Cartilage Fish
skeleton of cartilage
cold-blooded
paired fins

Jawless Fish
skeleton of cartilage
cold-blooded
jawless

Bony Fish
skeleton of bone
cold-blooded
gill covers

Amphibians
skeleton of bone
cold-blooded
gilled young; adults with lungs

Reptiles
skeleton of bone
cold-blooded
dry, scaly skin
leathery eggshell

Birds
skeleton of bone
warm-blooded
feathers
hollow bones

Mammals
skeleton of bone
warm-blooded
well-developed brain
feed milk to young

Circulatory Systems of Vertebrates (page 77)

Fish
two-chambered heart
gills
ventricle
atrium

Reptile
three-chambered heart
lungs
two atria
ventricle

Mammal
four-chambered heart
lungs
two atria
two ventricles

Characteristics of Fish (page 78)

Cartilage Fish
flexible skeletons
paired fins
5-7 gill slits per side

Examples:
skate
shark
stingray

Jawless Fish
sucker-shaped mouths
no paired fins
jawless

Examples:
hagfish
lamprey

Bony Fish
lobed- or ray-finned
gill covers
scaled bodies

Examples:
trout
moray eel
sea horse

Cartilage Fish (page 79)

1. kidney	2. vertebral column
3. stomach	4. heart
5. dorsal fins	6. caudal fin
7. liver	8. gill slit
9. eye	10. pelvic fin
11. intestine	12. pectoral fin

Bony Fish (page 80)

1. caudal fin	2. anal fin
3. pelvic fin	4. olfactory bulb
5. heart	6. liver
7. stomach	8. dorsal fin
9. dorsal fin	10. pectoral fin
11. brain	12. fin supports
13. swim bladder	14. anus
15. intestine	

Characteristics of Amphibians and Reptiles (page 81)

Amphibians
moist skin through which water can pass
adults breathe with lungs; young with gills
most adults are carnivorous
have three-chambered heart
have a cloaca
rarely have scales

Reptiles
tough, dry skin with horny scales
most hatch from eggs laid on land
have three-chambered heart
have a cloaca
young and adults breathe with lungs
eggs have a leathery shell

An Amphibian—The Frog (page 82)

1. nostril opening
2. mouth
3. foreleg
4. hind leg
5. eye
6. eardrum
7. nostril opening
8. teeth
9. tongue

The Internal Organs of a Frog (page 83)

1. lung
2. liver
3. kidney
4. anus
5. heart
6. stomach
7. mesentery
8. small intestine
9. large intestine

Life Cycle of an Amphibian (page 84)

1. frog is mature adult
2. an egg mass is laid in water
3. embryo grows within egg
4. egg hatches into larvae called tadpoles
5. tadpole grows hind limbs
6. tadpole frog has limbs and lungs and begins to absorb tail
7. young frog has absorbed all tail remnants

A Reptile—The Alligator (page 85)

1. kidney
2. testes
3. intestine
4. liver
5. lung
6. carotid artery
7. brain
8. posterior vena cava
9. cloaca
10. stomach
11. trachea
12. esophagus
13. heart

The Head of a Venomous Snake (page 86)

1. nostril
2. fang
3. pit
4. eye
5. venom sac
6. scales
7. glottis
8. teeth
9. tongue

Characteristics of Birds and Mammals (page 87)

Birds
have four-chambered hearts
have toothless, lightweight jaws
have wings
develop from embryos in eggs outside the mother's body
have feathers
have hollow bones

Mammals
have four-chambered hearts
raise young on milk
most give birth to live young
most have specialized teeth
have mammary glands
have hair

Parts of a Bird (page 88)

1. primary feathers
2. secondary feathers
3. tail feathers
4. back
5. crown
6. beak
7. throat
8. breast
9. belly
10. talons

A Bird's Internal Organs (page 89)

1. trachea
2. lung
3. wing bone
4. esophagus
5. eye
6. beak
7. spinal cord
8. kidney
9. gizzard
10. crop
11. heart
12. liver
13. sternum
14. rectum

Bird Beaks (page 90)

1. scoops up water and fish
2. cracks nuts and seeds
3. tears flesh of animals
4. sucks nectar from flowers
5. stabs small fish
6. traps insects in midair

Fowl Feet (page 91)

1. Function: for swimming
   Example: goose

2. Function: for catching prey
   Example: hawk

3. Function: for grasping in order to climb
   Example: woodpecker

4. Function: for wading in mud
   Example: coot

5. Function: for perching on branches
   Example: thrush

Birds of a Feather (page 92)

1. coverts	2. secondary flight feathers
3. primary flight feathers	4. body feathers
5. down feathers	6. primary flight feathers

Mammals (page 93)

1. monotreme	2. placental
3. placental	4. placental
5. marsupial	6. placental
7. marsupial	8. placental
9. monotreme	

Mammal Locomotion (page 94)

1. fly	2. swim
3. walk and run	4. swim
5. fly	6. walk and run
7. walk and run	8. walk and run
9. swim	

The Science of Ecology (page 95)

1. physical environment	2. biological environment
3. biosphere	4. ecosystem
5. Energy	6. food web
7. population	8. habitat
9. niche	10. Diversity

Producers, Consumers, and Decomposers (page 96)

1. decomposer	2. consumer
3. producer	4. consumer
5. consumer	6. producer
7. decomposer	8. consumer
9. producer	

Carnivores, Herbivores, and Omnivores (page 97)

1. omnivore	2. herbivore
3. herbivore	4. omnivore
5. omnivore	6. carnivore
7. carnivore	8. carnivore
9. omnivore	10. carnivore
11. herbivore	12. carnivore

Parasites (page 98)

1. attaches to fish	2. lives in intestine
3. enters the bloodstream	4. attaches to tree tissue
5. burrows into skin	6. lives off of other insects
7. blood-sucking insect	
8. blood-sucking segmented worms	
9. feeds off of forest plants	

Plant Succession (page 99)

See diagram on page 124.

Food Chains (page 100)

See diagram on page 125.

Food Webs (page 101)

See diagram on page 125.

Adaptations (page 102)

1. stems of plants store water in dry conditions
2. have special hemoglobin for high altitude living
3. hibernate to survive low food supplies in winter
4. have blubber to insulate body against cold
5. have specialized digestive tract to digest plant cellulose
6. have densely packed fur so skin does not get wet

Camouflage (page 103)

1. moth
2. walking stick
3. owl butterfly
4. caterpillar
5. snowshoe hare
6. walking leaf
7. halibut
8. deer

Defense Mechanisms (page 104)

Animal: tortoise
Defense: retreats into a shell

Animal: skunk
Defense: sprays offensive odor

Animal: octopus
Defense: releases a cloud of ink

Animal: ostrich
Defense: runs and kicks with powerful legs

Animal: opossum
Defense: pretends it is dead

Animal: porcupine
Defense: releases quills upon contact

The Deserts (page 105)

See diagram on page 126.

The Tundra (page 106)

See diagram on page 126.

Savannas and Grasslands (page 107)

American Grassland
contains grasses, shrubs, and few small trees
rainfall averages 10–40 inches (25–100 cm) a year
vegetation maintained by periodic fire
found between deserts and temperate forests
large grazing animals such as bison and antelope

African Savanna
contains grasses, shrubs, and few small trees
rainfall averages 35–60 inches (90–150 cm) a year
vegetation maintained by periodic fire
large grazing animals such as elephants, wildebeest, and zebra
found between deserts and tropical forests

Tropical Rain Forests (page 108)

1. more than half
2. nutrients
3. surface
4. microorganisms
5. vegetation
6. emergent
7. canopy
8. 2–5 percent
9. understory
10. forest floor

Coniferous Forests (page 109)

1. conifers
2. boreal
3. taiga
4. peat moss
5. pine barrens
6. temperate rain forest
7. montane
8. logging

Wetland Biomes (page 110)

1. salt marsh
2. marsh
3. impermanent
4. endangered
5. emergents
6. floating aquatics
7. swamp
8. bog
9. submergents
10. mangrove swamp
11. migration

Tidal Zone Biomes (page 111)

See diagram on page 127.

Coral Reefs (page 112)

See diagram on page 128.

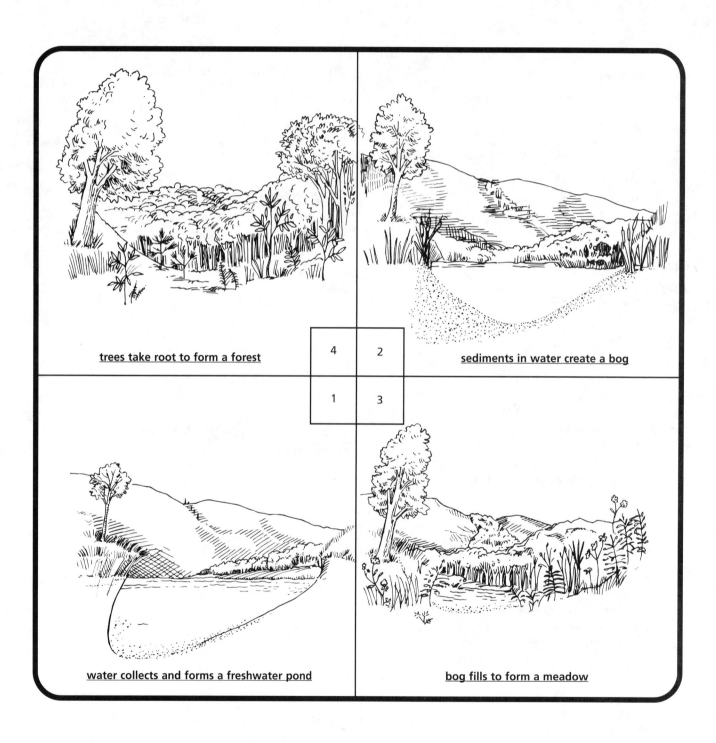

trees take root to form a forest

sediments in water create a bog

4	2
1	3

water collects and forms a freshwater pond

bog fills to form a meadow

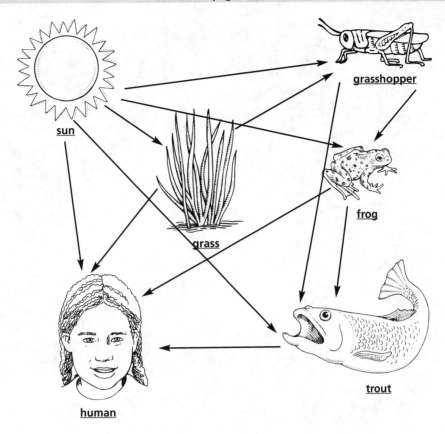

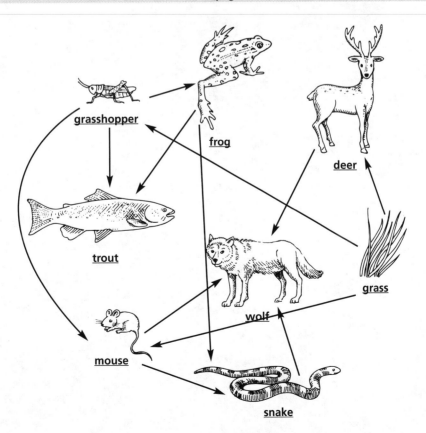

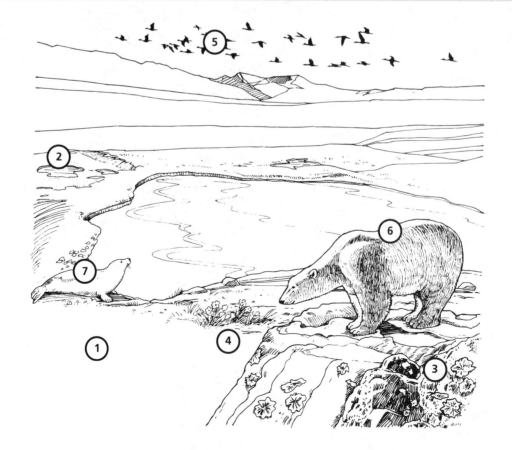

Zone: <u>upper intertidal splash zone</u>

Species: <u>barnacles and limpets</u>

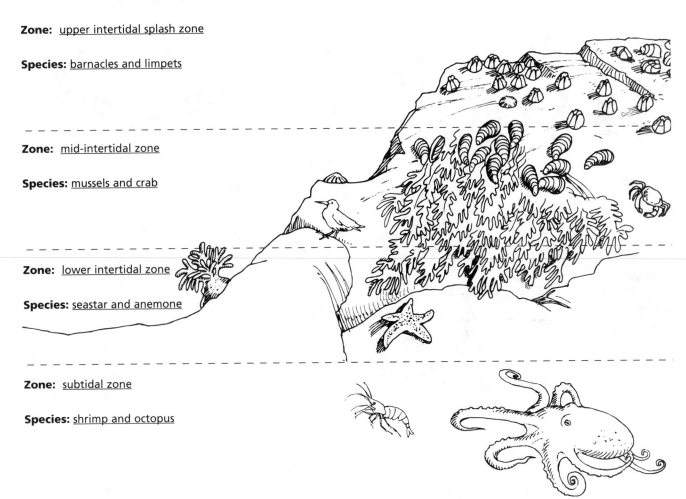

Zone: <u>mid-intertidal zone</u>

Species: <u>mussels and crab</u>

Zone: <u>lower intertidal zone</u>

Species: <u>seastar and anemone</u>

Zone: <u>subtidal zone</u>

Species: <u>shrimp and octopus</u>

fringing reef

barrier reef

lagoon

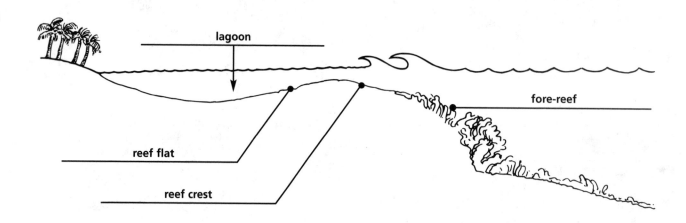

lagoon

fore-reef

reef flat

reef crest